INSIDE HBO

Inside
HBO

The Billion Dollar War between HBO, Hollywood, and The Home Video Revolution

George Mair

DODD, MEAD & COMPANY
New York

Copyright © 1988 by George Mair

All rights reserved

No part of this book may be reproduced in any form
without permission in writing from the publisher.
Published by Dodd, Mead & Company, Inc.
71 Fifth Avenue, New York, N.Y. 10003
Manufactured in the United States of America
Designed by Mike Cantalupo
First Edition

1 2 3 4 5 6 7 8 9 10

Library of Congress Cataloging-in-Publication Data

Mair, George, 1929–
Inside HBO

Includes index.
1. Home Box Office (Firm) 2. Cable television–United
States. 3. Moving-picture industry–United States.
I. Title
HE8700.72.U6M34   1988   384.55'47'0973   87-19939
ISBN 0-396-08420-6

Contents

What's It All About? ix
Acknowledgments xi
A Book Browser's Sampler of Quotations from the Text xiii
Introduction: The Billion-Dollar Giant xvii

I THE EARLY YEARS: 1972–78

Chapter 1. A Cash Cow is Born 3
Chapter 2. A New Industry in the Shadow
             of the Mountain 10
Chapter 3. The Father of Pay Television II 14
Chapter 4. Troubles from Inside and Outside
             for HBO 18
Chapter 5. Going Up on the Bird 22
Chapter 6. The Rush to the Sky 27
Chapter 7. Profits for HBO 31

II THE GOLDEN YEARS: 1979–83

Chapter 8. Cable TV—Promise or Phantom? 39
Chapter 9. The Infant Becomes a Giant 42
Chapter 10. HBO and Hollywood 46

Chapter 11.	Original Programming	51
Chapter 12.	The Money Side of Cable	55
Chapter 13.	1982—Excedrin Headache Year	60
Chapter 14.	*TV-Cable Week*	66
Chapter 15.	The Takeover Jitters	72
Chapter 16.	HBO Goes Hollywood and Hollywood Goes Cable	76

III CABLE TV ISSUES

Chapter 17.	T & A on Cable	83
Chapter 18.	What to Do about Home Video?	90
Chapter 19.	A Billion-Dollar Super Bowl and Jail for Pirates	95
Chapter 20.	Merger Mania	100
Chapter 21.	Programming, Marketing, and Ratings	105
Chapter 22.	HBO vs. Showtime	111

IV HBO AND TIME INC.'S TIME OF TROUBLES

Chapter 23.	HBO and Hollywood Try a Truce	119
Chapter 24.	The Warnings HBO Ignored	123
Chapter 25.	The Most Hated and Feared Man in Hollywood	127
Chapter 26.	Frank Biondi's Success and Failure	131
Chapter 27.	A New Hand at the Wheel	137
Chapter 28.	Inside the Executive Suites	139
Chapter 29.	Home Video vs. Pay-Per-View	142
Chapter 30.	The VCR Monster	145
Chapter 31.	Some Little Piggies to Market Wouldn't Go	155

Chapter 32. HBO Programming 1986–88   159
Chapter 33. Issues and Maybe Answers   167
Chapter 34. Cable TV's Giant as a Grown-Up 173
Chapter 35. Eating Broken Glass   177

Epilogue   188
Index   193

What's It All About?

Jerry Gross, my editor at Dodd, Mead & Company, like all good editors, insists that I tell him in one sentence the theme of this book. So, here goes.

Inside HBO is the story of an almost accidental company that became the biggest entertainment giant in the world in just 15 years.

A FEW WORDS AT THE START . . .

. . . There has never been an entertainment colossus to match Home Box Office. The business success of this young company is a story that has never before been publicly told. It is a company that skyrocketed from 325 customers in Wilkes-Barre, Pennsylvania, to more than 15 million customers all over the world ten years later.

Home Box Office is the single largest profit center for Time Inc.—a company that most people think of as a magazine publisher. Home Box Office is the largest financier and producer of motion picture entertainment. Its films are shown on pay cable television, in motion picture theaters, over satellite, and on home video VCRs.

I worked inside Home Box Office for almost a year as its Chief Public Relations Counsel. Since leaving HBO, I have spent two years researching, studying, interviewing, and analyzing HBO's success and its position in the entertainment industry and within Time Inc.

As a professional writer and journalist for over twenty-five years with CBS, the *Los Angeles Times* Syndicate, and others, I found the HBO story fascinating and untold. This book is the result.

George Mair
Los Angeles, California
1988

Acknowledgments

Thanks to . . .
. . . all the people who made this book possible:

Dr. Sandra Howell, my fiancée, who encouraged and supported me through the long hours the book took.

Jerry Gross, one of my Dodd, Mead & Company editors, who helped transform this book.

Allen Klots, another of my long-suffering editors at Dodd, Mead & Company, who tragically died before this book was published.

Lynne Lumsden, executive vice president and Publisher of Dodd, Mead & Company, who had the courage to proceed with this project when others wanted to abandon it.

Peter Osgood, president of Carl Byoir & Associates, who I sincerely believe was responsible for Carl Byoir's suing me in an attempt to frighten me into giving up this project before anyone at Carl Byoir or Home Box Office had even seen the manuscript. This made me determined to go ahead. It has cost me almost $20,000 in legal fees to fight these giant corporations for my First Amendment right to tell this story. The outcome is clear, because you are holding this book in your hands.

Anonymous people at Home Box Office and the National Cable TV Association who helped a lot, but whose jobs I would jeopardize by naming. They know who they are and so do I.

A Book Browser's Sampler of Quotations from the Text

Robert Lindsay of *The New York Times* says that the rise of Time Inc. to the number-one company in pay television is one of the major American business success stories of the last decade.

Home Box Office became the subsidiary of Time Inc. that soon accounted for almost half of Time's net profit. Yet, within a hundred days of having launched HBO for Time Inc., its creator, Chuck Dolan, was out. It's a tough business.

One unnamed cable subscriber expressed this feeling about what cable television had done for his life:

> "First, I saw the nude lady float down into the snow drift and begin wacking penguins and I thought, 'Wow, this is great.' Then, there were elephants—lots and lots of elephants—and I knew this is what television was really all about. That's why I love cable TV."

Another unnamed cable subscriber expressed a different view about cable television and Home Box Office:

> "I wish that King Kong would climb the Time-Life Building and break wind just as he passes your office, you scum."

xiv *Quotations from the Text*

Dr. Walter Baer of the Rand Corporation says that, after years of a glorious development, the cable TV industry appears to be just starting its childhood.

In the parking lot of the Disneyland Hotel, Jerry Levin—the man who invented pay TV the second time around—saw the future. It was his future, HBO's future, and television's future.

Cable TV is filled with promises and phantoms. Is all the profit that cable TV is going to make a promise or is it a phantom? The answer is "yes."

Jack Curry, writing in the New York Daily News, once said that, if you showed him a man who subscribed to HBO, he would show you a man who had seen *The Great Santini* fifteen times.

HBO and Hollywood have the symbiotic relationship of mutual parasites who loathe each other but couldn't survive without each other. To use an old Texas expression, they are like two strange dogs on their first meeting, circling and sniffing, sniffing and circling. Each one would have no greater pleasure than to lift his leg on the other.

Time Inc.'s Video Group brings in 19 percent of the company's gross profit and 42 percent of its net profit. This is from a business that didn't exist ten years earlier.

In home video, the Japanese once again took an American invention and turned it into a gold mine while American industry sat around sucking its thumb. Today there is not a single VCR made in America. All VCRs are made overseas, mostly in Japan and Korea.

Former Dallas Cowboys quarterback Roger Staubach has been warning people around the country that cable TV would bring homosexual acts, women being brutally molested and raped, and other explicit sex acts into their living rooms.

It is clear that the VCR is doing to the pay cable business what the pay cable business did to broadcasting thirty years before. The spread of VCRs and home video in America has been explosive and is drastically changing both the television and advertising businesses.

Nick Nicholas, the then head of Time Inc.'s Video Group, told a shocked audience of cable TV executives that what HBO and cable had been doing for the last few years was a fraud on the consumer.

Esquire magazine has called Michael Fuchs, now Chairman of HBO, "the most potent, feared and hated man in Hollywood."

A survey of what is important to married couples conducted by Dr. Sol Gordon and Dr. Kateryn Everly ranked money #1; HBO #2; and simultaneous orgasms #3.

In 1981, HBO became Hollywood's biggest single customer, and in the next ten years, HBO will grow an estimated 400 percent.

Introduction:
The Billion-Dollar Giant

In the history of the entertainment industry, there has never been such a supergiant as Home Box Office. This book will tell you how it got to be that way.

Home Box Office (usually called HBO) is a division of Time Inc. Most people think of Time Inc. as primarily a magazine publishing company, but it isn't. The number-one source of Time Inc.'s net profit—46 percent of it—is from television; magazines bring in 36 percent.

It was not always so. Back in 1973 the television operation was a virtual nothing, and in fact Time Inc. tried hard to get rid of it, but was forced to keep it by the New York City Council which refused to approve the sale.

Today Time Inc. is the second-largest cable TV owner in the world through its subsidiary, American Television and Communications (ATC), which is based in Denver. It is also the largest TV programmer in the world through its enormously profitable subsidiary, HBO.

The success of HBO is one of the great, untold business stories of the last decade. Its fabulous success is due to a handful of smart men and women at HBO, a lot of stupidity by its competition, and some good old-fashioned luck.

In becoming a success, HBO had to challenge the old-time movie moguls of Hollywood who have been in control of much of American entertainment for years. In 1973, HBO was coming hatin-hand begging Hollywood for some cooperation. Today, Hollywood is terrified that HBO will soon take over the world.

Today, HBO is the world's largest single financier of movies. Starting as a movie middleman between the cable system owners and the movie studios, it is now involved in movie production, pay cable programming, concert promotion, theater exhibition, original productions, even home video.

When you see a movie produced by HBO or by Tri-Star or by Silver Screen Partners or by Orion, you are seeing a movie that involves Home Box Office. In addition, many of the other movies you see from the other studios may have been partially financed by HBO. HBO is involved in more movies than four of the big seven Hollywood studios combined.

This is one of the reasons that Barry Diller, former head of Paramount Studios and now with Fox Broadcasting, predicted that HBO would control Hollywood in the next few years unless somebody found a way to stop it. In fact, almost every studio *has* tried to stop the explosive growth and power of HBO, but they have failed.

Every month over 15 million Americans spend an estimated ten dollars to bring Home Box Office into their homes. That's $150 million a month in cash—not credit—$1.8 billion every year. Is it any wonder that HBO is regarded as The Cash Cow That Almost Ate Hollywood?

HBO was basically the vision of two men: Chuck Dolan, who created the idea, and Jerry Levin, who made the idea work. The first five years (1973-1978) were a struggle, and the next five years were a dream, with HBO making more profit than anybody—inside Time Inc. or outside—ever thought possible.

Then the giant stumbled. In 1983-85, Time Inc. management made a lot of mistakes and so did Home Box Office. Even so, HBO never lost its dominance of pay cable TV. It made *less* profit than before, but it still made a substantial profit for Time Inc.

Now the HBO giant is recovering his balance, if not his complete self-confidence. In 1987, it continues to be the colossus of the entertainment world in the United States and is expanding overseas. It is still the most profitable operation Time Inc. has going, and it is likely to get even more profitable in the future.

This is the story of how it all happened.

INSIDE HBO

PART I

THE EARLY YEARS: 1972–78

Chapter 1

A Cash Cow is Born

It all started in 1971 when Chuck Dolan was in trouble.

Dolan is a hustler in a business made up of hustlers—cable television. He is a successful hustler because he is action-oriented and imaginative. In 1965, he founded a small cable company, Sterling Communications, as a subsidiary of his company, Sterling Information Services, and got the exclusive franchise to provide cable service to lower Manhattan.

In 1965, Time Inc. had made what it considered a high-risk investment of modest size by buying 20 percent of Sterling Communications for $1,250,000. Cable TV was an uncertain medium with scant financial history on which to base any sound analysis; still, Time Inc. made the investment. This investment was even more difficult to understand in light of the general inclination of Time Inc. to get out of the electronic media business. It owned five television stations and was casting around for a buyer for them.

What Time Inc. executives soon learned about cable TV was that it had an enormous appetite for cash. This is particularly true if one is building a cable system in a metropolitan area as Sterling was doing. In the country or suburbs, cable can be strung quickly and relatively inexpensively for about $10,000 a mile. However, in cities where cables must be taken underground through a bewildering maze of tunnels, tubes, and ducts in order to reach the customer, the cost is ten to thirty *times* as much. So, by the middle of 1967, Sterling had spent $2 million and had only reached four hundred customers.

Executives at Time Inc. estimated it would cost Sterling at least another $10 million to complete its wiring of lower Manhattan. The critical question of where it was going to get that kind of money arose. The obvious answer was a bank, but banks weren't enthusiastic about lending to cable companies in those days.

Sterling had gone to the Chase Manhattan Bank and the bank had agreed to make a $10 million loan, but only under rigid conditions. First, all the individual stockholders in Sterling Communications would have to guarantee the loan. Chuck Dolan was willing; Time Inc. was willing; but Dolan's two partners, Elroy McCaw and William Lear, were not. This problem was finally solved by McCaw and Lear's selling out to Time Inc. thus giving it almost half ownership of Sterling Communications.

The other proviso set down by the bank was that it would release only half of the money right away. The remaining five million dollars would only be released as certain customer levels were reached, with the first level being 12,500 subscribers. That took until July of 1969.

Meanwhile, to tide Sterling Communications over dry spells in its cash flow, Time Inc. began making short-term loans to the cable company. However, by the fall of 1969, it was clear that the cable company was gobbling up money faster than anybody had anticipated and Time Inc. said that it would only lend more money if the funds were convertible into a bigger share of ownership.

It was New Year's Eve when Dolan and the people at Time Inc. struck a deal at last. Dolan would give up 44.5 percent of the parent company, Sterling Information Services, in exchange for continued loans from Time Inc. and these loans would be convertible into more equity in the parent company.

Even so, the people in the executive suites at Time Inc. were taking a hard look at all the nonpublishing activities of the company. There was a growing feeling that the cable TV business was the wrong business at the wrong time. However, Barry Zorthian, the new head of the Time-Life Broadcast Division, said he had faith in cable, and his assistant, Edgar Smith, predicted that cable would someday be a profitable industry.

Naturally, the company didn't make a profit at the beginning; four years later, it was still losing lots of money. Dolan was casting about for some way to turn things around financially. Even so, the pressure was on Chuck Dolan to produce a winner.

So it was that, on a vacation to Europe with his family aboard the ocean liner Queen Elizabeth II, Dolan came up with a plan. He wrote out a proposal for what he called the Green Channel: a subscription pay channel focusing on sports and movies.

A Cash Cow is Born 5

As Dolan envisioned it, the Green Channel would rent movies from the Hollywood studios, just as movie theater houses did, and show selected sports events. Naturally, this would be too expensive for the Green Channel to afford by itself, so Sterling Cable would get other cable TV system owners to let Sterling operate the Green Channel programs on their cable systems.

By spreading the cost of movies and sports events this way, Sterling could afford them; and the programs, in turn, would attract more cable subscribers.

A few days after getting back from his vacation, Chuck took his idea to Barry Zorthian. Zorthian, a bearlike hulk of a man, was a longtime Time executive with a lot of news experience. (He would later spend a tour of duty in Vietnam and finally end up as head of *Time* magazine's Washington, D.C., news bureau.)

Zorthian wasn't completely sold on Dolan's proposal—he thought it had some weaknesses—but he sensed that the Green Channel was something they had to try.

What made the Dolan vision and the Zorthian decision especially daring was the climate of the times. First of all, cable TV was the new kid on the electronic block and was under assault by the powerful TV broadcasting networks and the Federal Communications Commission (FCC). Second, a lot of executives at Time Inc.—particularly on the magazine side of the business—weren't sold on television or cable TV. Even so, Zorthian was willing to commit $150,000 to the development of the Green Channel idea. On November 2, 1971, the Board of Directors of Sterling Communications approved the Green Channel plan.

With the backing of Time's seed money Dolan put together his Green Channel start-up team. The first to come on board was a lawyer, Frank Randolph, and a well-known New York sportscaster, Marty Glickman, who would get HBO contacts and contracts with professional sports teams. Then Dolan hired Tony Thompson, who came over from Time Inc. to handle the marketing of the new channel, and another lawyer, Jerry Levin, whom Dolan and Thompson had met at a cable association convention in Chicago and hired in early 1972.

Levin, thirty-three, had started out as a divinity student and then became an antitrust lawyer with the Wall Street firm of Simpson, Thacher and Barlett. After Wall Street, he later worked for the Development and Resources Corporation, which did international consulting. In that job, he created a carnation nursery in Colombia and supervised the building of a major irrigation system in Iran. He understood how to marshal money and people.

Levin would prove to be Dolan's smartest pick. He was a good administrator and a sports and movie nut. He has been described as "a marvelous negotiator with a good legal mind." It was Levin who eventually sold the concept of Home Box Office to the doubting magazine executives who ran Time Inc.—the ones who didn't believe people would pay for TV when they were already getting it free.

Early in the project's development the team of Dolan, Levin, and Thompson decided to change the name of the programming service from the Green Channel to something more appropriate. After kicking a number of alternatives around, they "temporarily" picked the name Home Box Office under pressure of a brochure printing deadline. They never got around to picking a permanent name.

Soon after joining forces, Levin, Dolan, and Thompson decided to conduct a market survey to see if people would like the concept of HBO. There wasn't much point, they agreed, in trying to launch HBO if nobody wanted it enough to pay for it. Naturally, they were depressed when only 1.2 percent of the people surveyed thought it was a good idea.

Discouraged, but still not dissuaded, they did a door-to-door survey in four cities, explaining HBO face-to-face. The result this time was that 50 percent of the people interviewed said they would subscribe. That's when they knew they had something.

The HBO trio faced another problem with film suppliers. Before the suppliers would rent their films to HBO, they wanted to know how big an audience HBO had and how much it would pay. Movie studios were used to getting a piece of the ticket price paid by each customer of the movie theaters. Their mind-set prompted them to try for the same system of payment from this new kind of at-home, electronic movie theater. The only problem was that no one would be selling tickets.

Cable system operators also had their doubts about the new venture. Before they would sign over use of one of their channels to HBO, they wanted to know what kind of movies would be on and how much HBO was going to pay for them. Certainly, no cable operators wanted junky or offensive movies showing on their cable system; the operators feared that their subscribers would be driven away. Beyond that, the cable operators wanted to pay as little as possible to Home Box Office.

The deal that Home Box Office wanted and the one it was finally able to make with movie studios and cable operators both was to pay the studios a flat fee rental and to make the cable operators participating partners in the HBO service on their cable systems. This would give the cable system operators an incentive to promote HBO, because their income depended on the number of basic-cable subscribers who pay the extra monthly fee to get the premium service of Home Box Office.

To get the ball rolling and a cable system signed up, Chuck Dolan came up with an inspired idea. Instead of trying to negotiate with a single cable operator for the exclusive franchise for one particular city, he looked for a city with two competing cable systems. He found that in Allentown, Pennsylvania, which had two cable systems instead of the usual single monopoly. He played one against the other and ended up with both of them wanting HBO.

While Dolan was signing up HBO's first cable system, Marty Glickman and Jerry Levin were busy lining up programming. Glickman got Madison Square Garden to agree to let HBO televise National Basketball Association (NBA) games live from the Garden and Levin persuaded Universal Pictures to rent some of its films.

Soon thereafter, Dolan gathered his team and announced, "Now, we're ready to go." Unfortunately, that was not quite true. There was still a stumbling block: a man named Ed Snyder. Snyder, who ran the Philadelphia Spectrum Arena, had the exclusive rights to all NBA broadcasts within seventy-five miles of Philadelphia. Allentown fell within that seventy-five-mile circle, which meant that NBA games could not be shown except by Snyder. HBO was frozen out.

That's when luck, in the shape of John Walson, stepped in to save HBO. Walson owned the Allentown system HBO had signed up as its first pay TV partner, but he also owned another system outside Snyder's seventy-five-mile circle. He offered to put HBO on that system, but warned Dolan that there was one small problem.

The problem was that the little community, Wilkes-Barre, Pennsylvania, was virtually under water as a result of Hurricane Agnes. Even so, Dolan jumped at the offer and sent Thompson down to get things rolling. Walson had not been kidding about the hurricane. In the store that Thompson rented for the HBO office the waterline was clearly visible six feet up the wall.

Tony and his crew cleaned up the place and started going door to door selling HBO. While Tony's crew was signing up subscribers, HBO's chief engineer, Bob Tenten, was getting the technical side built. The programming would originate in New York and be transmitted by microwave towers to Wilkes-Barre. Then it would go over the cable TV system's lines to subscribers' homes.

The target date for launching the first programming over HBO was the evening of November 8, 1972. Just before the launch, Time Inc. shifted a new player to the top HBO slot, J. Richard Munro. A decorated ex-marine who had seen combat in Korea, Munro had been publisher of *Sports Illustrated* magazine.

Munro's assessment of his new assignment was that it was "a crazy business where nobody seemed to know what was going on." (Munro would be at HBO only a few months before he was replaced by Jerry Levin. Munro would continue up the ladder until he became the top executive at Time Inc.)

Early on the day of the HBO launch, Munro decided that he and Tony Thompson ought to be in Wilkes-Barre for the opening ceremony, but they got stuck in a monumental traffic jam on the George Washington Bridge leaving Manhattan. They called Wilkes-Barre and told the rest of the group that they couldn't make it to the ceremony. Not to worry, they were told. Nobody in town thought it was a big deal anyhow. The local newspaper had decided it was not worth covering and the mayor had canceled his appearance.

Besides, the HBO people in Wilkes-Barre had more important problems. The weather was bad, and a high wind had knocked down their microwave receiving dish. The technical crew wasn't able to get their repair work finished in time for the first HBO program to go on the air. HBO began its first programming with the receiving dish on the roof in Wilkes-Barre being held in place by a member of the technical crew. Prophetically, the first movie shown on HBO was *Sometimes a Great Notion*. During that first month, HBO showed nine NBA games, five National Hockey League (NHL) games, and four movies.

Within a hundred days after HBO began in that small Pennsylvania town, Chuck Dolan was out of Home Box Office. To this day Dolan and the people at HBO refuse to talk about why, but

Dolan was to continue in the business with his own company, Cablevision, which operates cable systems in places around the country such as Long Island, Connecticut, and Boston.

It's a tough business.

Chapter 2

A New Industry in the Shadow of the Mountain

In 1928, television, the video disk, and medicated socks were all invented by John Logle Baird. None of his ideas blossomed right away, so Baird, an eccentric Scotsman, had to make his living canning jams and making boot polish in England.

Just before the beginning of World War II, there were about ten thousand TV sets in the United States. Naturally, they bore little resemblance to the ones we have today; most had tiny three- or four-inch screens. By the late 1940s, however, television was booming in America, but it was a hit-or-miss proposition. Some people lived in areas where mountains were between them and the closest TV transmitter; they got very poor pictures or no pictures at all.

The reason for this problem is simple. Television signals are transmitted by FM waves, which travel in a straight line. (AM radio waves have the ability to bounce off the bottom of the ionosphere and then wrap around mountain ridges and tall buildings, which is why you can hear AM radio stations in places you cannot hear FM radio stations.) Television transmitters are usually mounted on the highest place in the community so that they can travel in their straight lines directly to your antenna. However, if a mountain or even a tall building is between your TV antenna and the TV transmitter, you will get either no picture or one that has a lot of ghosts. In the early days of television there were many communities screened from their television transmitters in this way.

A New Industry in the Shadow of the Mountain

Mahanoy City, Pennsylvania, was one of those places. The picture from the Philadelphia TV stations eighty-six miles away was extremely poor because of the mountains in between. John Walson had an appliance store in Mahanoy City, but he had trouble selling his TV sets because of the poor reception in the area. As luck would have it, Walson also had a job as a lineman for the Pennsylvania Power and Light Company. One day in 1948, the enterprising Walson put an antenna on the top of a nearby mountain and ran cables down to the town. In short, John Walson invented cable TV.

He called his company Service Electric Cable TV, Inc., and charged his customers $100 for the initial hookup and the first year's subscription. Thereafter, cable service cost was $2 a month.

On Thanksgiving Day that same year in Astoria, Oregon, Ed Parsons rigged a community antenna on top of the Astor Hotel and picked up KRSC-TV a hundred miles away in Seattle, Washington. Then he ran a cable from the hotel roof to a TV set in the local music store, Cliff Poole's, so that people could gather around and watch television. Cable TV was spreading.

The idea of a master antenna on a high place connected with local TV sets by a wire or cable began to spread around the country. In 1952, four years after Walson first created his cable television system, there were seventy such systems, with fourteen thousand subscribers. While Chuck Dolan was writing his proposal for the Green Channel in 1971 aboard the *QE II*, the industry had grown to 2,600 cable systems and 5.3 million subscribers.

By the time Chuck Dolan had left the fledgling HBO, in early 1973, the original 325 Wilkes-Barre subscribers had grown to 1,537. However, during that same time, HBO had also *lost* 1,229 subscribers. In other words, for every one subscriber HBO netted, it had to sign up 1.8 subscribers.

Signing up for a service and then dropping it a short time later is called *churning*, and it is a problem that has plagued cable TV from its earliest days. Yet, even with churning, HBO was growing. Besides its 1,537 subscribers in Wilkes-Barre, it had added almost 4,000 more in nearby Bethlehem and Allentown. HBO was unquestionably a success.

Much of that success was due to four management decisions made at the beginning. First, HBO decided to charge a monthly flat fee for service rather than charging for each program a subscriber watched. This eliminated what would have been an enormously complicated bookkeeping problem. (This problem would show up again in a few years when some people in the cable industry decided to charge for each program under a system that would be called *pay-per-view* (PPV). We'll talk more about that in a later chapter.) Besides, the magazine-oriented executives at Time Inc. related better to a monthly charge system because that's how their magazine subscriptions worked.

Second, the cable operators became "partners" of HBO because they shared in the revenue-producing incentives to sell HBO to their basic subscribers. HBO also helped the cable operators promote HBO sign-ups by providing them with advertising and promotional material to send to present subscribers and for door-to-door salesmen to use in getting new subscribers.

In a subtle way, however, the use of HBO promotional materials by a cable system's sales force might have had a long-term negative effect. The representative would emphasize the great blockbuster movies the subscriber would get by signing up for basic cable (you have to subscribe to basic cable before you can get any premium pay services such as HBO) and HBO, but unfortunately, there weren't that many blockbuster movies available on HBO and those that *were* available were repeated over and over again. Too often the customer meant to sign up to see *The Great Santini, Terms of Endearment,* and *Raiders of The Lost Ark* when, in fact, all that was shown were *The Attack of The Killer Tomatoes* and *The Amazing Colossal Man.* Disappointed because of the overzealous sales pitch, the subscriber would get angry and order that the cable service be disconnected. (Churning continues to be a cable television problem even now, in 1988.)

Third, HBO decided to transmit its programs electronically, by microwave instead of shipping movies and tapes by mail. This turned out to be a very important decision, because it made HBO attuned to the latest technology right from the start. As we will see, by staying on the leading edge of technology, HBO would continue to be successful and would virtually re-invent pay television.

And fourth, HBO settled on a sports-specials-movie mix of programs. Cable TV has never been able to get enough good movies, so this mix of programs helped round out the offerings. This variety would soon shift HBO's market position from that of a program broker operating between program producers and cable systems to that of a program producer itself. This shift in market position, done in some cases inadvertently in the search for more programs, would enable HBO to defend its position in an increasingly competitive business.

You can perhaps better understand HBO's problem of not having enough programs when you realize that all of the studios in Hollywood turn out only about eighty to one hundred movies a year. If each runs about two hours, that's less than two hundred hours of programming a year. A pay TV service such as HBO has 8,760 hours a year to fill. Even if you factor in the several thousand old movies in various film libraries and assume you could get the rights to show them, there just aren't enough old movies to fill all the time on the four major pay TV networks that now exist. It's no wonder the same movie is shown several times a week.

Besides buying the cable TV rights to every movie it can, HBO sponsors special programs and underwrites sporting events. In addition, HBO has been involved in the financing and production of movies since early in its existence. That may solve one problem, but it creates another. HBO is in conflict with the movie studio moguls, who see HBO as an interloper invading their territory.

Like most business conflicts, the battle between HBO and Hollywood is simply over money—in this case, billions of dollars. It is a battle that the studios are destined to lose. Prominent among the reasons for the studios' being weak when fighting HBO are two that seem to be inherent in the nature of Hollywood movie moguls.

One, Hollywood has never grasped the importance of new technology. It fought against the talkies, against color, against radio, against television, against cable TV, and, most recently, against the videocassette and home video. Hollywood ended up profiting from all of these new technologies, but only after it tried unsuccessfully to kill them. Hollywood always views a new way of doing things as a threat, instead of as an opportunity.

Two, Hollywood movie executives are notorious in their dislike for one another. Their rivalries keep them from joining forces to deal with interlopers such as Home Box Office. One Hollywood observer says that each movie producer would rather see another producer go down the tube than to be forced to work with him.

Still, in spite of all the carping and complaining Hollywood does about the power and arrogance of HBO, the new industry pumps millions of dollars into Hollywood pockets every year. HBO was to become the largest financier of motion pictures in the world by 1987, and was then involved with many production companies. It was destined to be the biggest cash cow Time Inc. had ever had.

Chapter 3

The Father of Pay Television II

Just about five months after HBO started in business, in November of 1972, Jerry Levin took charge as president. To some, it looked as if he had become captain of the *Titanic*.

A short man with a bushy mustache, Jerry Levin had an interesting background. He had studied to be a priest, switched to law, built construction projects in the Middle East, and ended up being one of the most important people in the history of television and HBO.

It is said that every successful business needs a dreamer, a businessman, and a son of a bitch. HBO had all three. Chuck Dolan had been the dreamer and Jerry Levin was the businessman. Some people in Hollywood say the current Chairman of the HBO Board, Michael Fuchs, is the s.o.b., but others are more charitable.

As HBO was starting under Levin, cable TV was beginning to take on its modern form, moving away from the mom-and-pop operation that characterized its first twenty-five years. As it began to mature, getting enough programs became an even bigger problem.

HBO's growth in its first year was uneven. As previously mentioned, it had expanded to Bethlehem and Allentown and added almost 4,000 subscribers to its original 1,537 from Wilkes-Barre. But then by the end of 1973, HBO's first full calendar year, the number of subscribers had fallen from a high of 12,500 to 8,323.

There are at least three reasons for the drop in HBO subscribers. One, in the summertime, many people spend more time outdoors and less time watching television. When they come back to TV, it is easier to start watching over-the-air, free television than it is to connect the cable. Two, HBO had unwisely depended on the cable operators to sell HBO. Cable operators, particularly during this era, were not merchandisers or salespeople. They were mostly engineering types. Thus HBO was not properly marketed. Three—and this was the most bothersome—many subscribers did not feel they got their money's worth by subscribing to HBO.

Levin decided that his own sales crew needed to go back into the Wilkes-Barre system and resell HBO. He remembered that HBO was best sold in person, door to door, so in went the sales crew and the effort paid off. During the month of December 1973, for the first time in almost a year, HBO signed up more subscribers than it lost.

A new marketing man, Bill Myers, came over to HBO from *Sports Illustrated* to get sales back on track. By the spring of 1974, HBO had fifteen thousand subscribers. That was a marked improvement, but it still didn't satisfy Levin. He knew that the top executives at Time Inc. were not convinced that cable was a viable business or that they should be in it. Levin knew that if he didn't show some dramatic results fairly quickly, HBO's life expectancy would be relatively short.

Levin demanded that the staff get twenty thousand subscribers by June, which meant netting five thousand new customers in two months. That was a pace that had never before been reached at HBO, but Levin was under a lot of pressure from Time Inc. president Jim Shepley to reach that twenty thousand mark by mid-year and forty thousand by the end of the year. In his memo to aides Jim Heyworth, John Barrington, and Stuart Chuzmir, Levin made his position clear: "As of June 30: the magic 20,000. There are no ands, ifs, or buts about it."

One of the keys to selling HBO was having good movies, but there were restrictions to cope with. For one, the FCC wouldn't let HBO or other pay TV firms rent movies unless the movies were less than two or more than ten years old.

In the 1970s it was tough to get the studios to rent new movies less than two years old to HBO, because the studios generally kept movies on the theater circuit for a couple of years. First, the studios would release them to the big urban theaters, which had lots of promotion and high ticket prices. Then it would release them some months later to the neighborhood theaters, and finally, to what are called the second-run theaters. This scheduling generally ruled out new movies for HBO.

In fact, some film studios, such as Walt Disney, wouldn't rent *any* movies to pay TV, much less new ones. The only break in the situation came when Paramount agreed to release *Charlotte's Web* to HBO, but that caused a big flap with the theater owners. The irony of both these instances is that Disney would later get very heavily involved with cable TV itself, and Paramount, which helped HBO out in these early years, would deliberately try to hurt HBO a few years down the road.

HBO got another break in 1973 from the Organization of Petroleum Exporting Countries (OPEC). When OPEC imposed an oil embargo on the United States and gasoline prices skyrocketed, many people stayed home and turned to TV for their entertainment.

HBO capitalized on the situation by instituting "Gasless Saturdays and Sundays" and expanding its programming on the weekends. In that year, Levin boosted HBO programming hours from an average of four hours a day to seven and a half hours a day. Levin hoped that the gimmick would attract more customers, but there was a downside to this plan: It meant finding more programming to fill those extra hours.

Levin was a man pushed. He was pushed by his own driving energy; he was pushed by Jim Shepley, his boss at Time Inc.; and he was pushed by suppliers of films and sports programs, who wouldn't give him enough programming.

So, Levin made a bold move. He told the National Basketball Association (NBA) and the National Hockey League (NHL) that he was going to televise their games whenever their teams weren't playing at home and there was no conflict with some other local television station showing the game. In other words, he would show his HBO subscribers games that nobody else would see anyhow. And, he wasn't going to pay a dime for the privilege.

The reaction from the NBA and NHL was a lot of grumbling and grousing, but surprisingly, they decided not to fight with HBO. This has to be regarded as a little odd, because here was Levin saying he was going to take programming the NBA and NHL had created and show it for pay to his customers without paying the program's creators. Still, that's what happened.

Next, Levin made a deal with WPIX, the independent New York TV station, about Yankee baseball games. WPIX had the rights to show all the Yankee games, but the station was televising only about half of them. Levin's proposition: "What you don't show, let us show." WPIX agreed to the deal. Broadcasters complained to the FCC, but the FCC shunted the complaints aside.

Jerry Levin was beginning to make HBO and pay television work. What he needed was some time. He was granted time and went on to become known as The Father of Pay Television II. But his story continues, as we will see later.

Chapter 4

Troubles from Inside and Outside for HBO

HBO was not the first pay cable TV service in the industry. There had been two other significant attempts at creating a pay television system. One was tried in Hartford, Connecticut, but it died after a few years of testing. The other, a more ambitious venture, was started in California by a former vice president of NBC, Pat Weaver, who had created other television landmarks: "The Tonight Show" and "The Today Show." Upon quitting NBC, Weaver moved to California to start a pay TV service and was immediately under siege by the movie theater owners, who succeeded in getting a proposition on the ballot forbidding pay TV.

A lot of legal scholars said the ballot proposition was unconstitutional because the law cannot forbid anybody to operate a legitimate business. After all, it was not as if pay television was corrupting the public morals. The proposition was allowed to stay on the ballot, and, after a vigorous campaign, in which the theater owners railed against the public's being charged for something they now got free, it passed.

Weaver and his company went to court to challenge the law, and finally, several years later, the California Supreme Court ruled the law unconstitutional. By that time, however, Weaver's money, energy, and ambition were exhausted. He did not go ahead with his plan.

In the mid-1970s, just as it began to crystallize into a major business, cable was attacked on all sides as an unwelcome competitor for America's entertainment dollars. It also had an image problem because of scandals in the young industry. For example, Irving Kahn, the president of one of the largest cable companies, TelePrompter, was convicted of conspiracy, bribery, and perjury in connection with TelePrompter's bid to get a cable franchise in Johnstown, Pennsylvania. He went to jail for five years.

In addition, the Federal Communications Commission (FCC), the federal agency charged with regulating broadcasting, was a recurrent source of trouble to the cable industry. It seemed to be trying to protect the broadcasting industry against cable TV. Like the broadcasting industry, the FCC didn't seem to know what to do with cable. First it said it wouldn't regulate it at all because it wasn't broadcasting. But just as the broadcasters decided that cable TV was an economic threat, the FCC reversed itself and discovered that it could and would regulate cable TV.

In June of 1972, right after Levin joined HBO, the TV network's trade association, the National Association of Broadcasters (NAB) pushed for tighter cable regulations, and the FCC gave them what they asked for. The FCC ruled that cable couldn't show any sports event that had been on commercial television until five years after the fact.

Broadcasting also was doing everything it could to stifle cable TV's development. For example, a vice president of ABC-TV, E. Erlick, testified that when ABC bought a film from a movie studio, ABC stipulated that the film must be kept off pay TV.

The commercial TV broadcasters were not the only ones up in arms about cable; the movie theater owners were worried, too. In August of 1972, three months before HBO was launched, Sherrill C. Corwin, a director of the National Association of Theater Owners (NATO), realized that cable was big business. He warned his colleagues about the dangers of cable TV and also urged them to get into cable themselves.

They didn't listen to him. Rather, the theater owners put pressure on the movie studios not to rent movies to HBO. Since the theater owners were the biggest customers of the studios, this was a serious roadblock for Jerry Levin and the struggling HBO to overcome.

Just before Levin took over as president of HBO, *Daily Variety* reported that the theater owners had organized a task force to fight cable TV in Congress. The task force had the cumbersome name of "The Committee to Protect the Public from Paying for What It Now Gets Free."

If the movie business was considered tough for HBO to enter, things were not easy with sports programming, either. Just as theater owners were determined to resist cable TV, the sports broadcasting interests were prepared to put up a fight. Jack Schneider, president of the CBS Broadcast Group, warned the U.S. Senate Commerce Committee that cable TV was a menace and asked the Committee to forbid the sale of professional sporting events to cable. Baseball Commissioner Bowie Kuhn echoed Schneider's position. (This reaction is reminiscent of the approach taken by some professional sports team owners who wanted the Congress to protect their sheltered monopoly on the grounds that professional sports was not a business but a casual recreation.)

But there was some good news. Some sources viewed cable as a great communications breakthrough. The Rand Corporation, a California think tank, issued a study predicting that cable TV would influence our lives in the future as radically as the automobile and the telephone had in the past.

Still, Time Inc. was not into cable because it was enthused about the future of an America wired to cable TV. Time Inc. was a business, and they expected their business to make money. And they really were not too sure about the broadcasting and the cable business.

Rhett Austell, vice president of the TV & Film Group of Time Inc., warned that the company's television activities were going to take a lot more money and that growth would be much slower than originally anticipated.

Meanwhile, Time Inc. was getting deeper into the area that had originally spawned HBO, Sterling Communications, even though Sterling's financial situation continued to get worse. Sterling had lost $2.5 million in 1971 and $4 million in 1972. Time Inc. continued to pour money into the cable operation, and it was converting its loans into a bigger and bigger ownership share.

By March of 1973, Time Inc. owned 66 percent of the parent company, Sterling Information Services, and when September arrived, the share had increased to 79 percent. The executives at Time Inc. decided one of its problems at Sterling was Chuck Dolan. So Dolan was shuffled out as head of Sterling Communications and Dick Galkin was put in as president. At HBO, Dolan had been replaced, as we saw, by Jerry Levin.

Then, Time Inc. bought out Dolan completely, dissolved Sterling, and made both HBO and the cable company (renamed Manhattan Cable) subsidiaries of Time Inc.

Troubles from Inside and Outside for HBO

Although it was getting deeper into Sterling and HBO at that moment, Time Inc. decided to get out of the cable system business entirely. It had already sold off its radio and TV stations around the country, and on May 8, 1973, Time Inc. made a deal with Warner Cable to sell the ten cable TV systems it operated under the name Time-Life Cable Communications.

At that time, Time Inc. was also a 9 percent stockholder in a large multisystem cable operator, the American Television and Communications company based in Denver. In the fall of 1973, Time decided to get out of that, and it sold its share.

Six weeks after Jerry Levin became president of HBO, Time Inc. agreed to sell Sterling Manhattan Cable—from which Levin and HBO had sprung—to Warner Cable for $20 million.

A few weeks later the Warner deal fell through when Warner discovered that Sterling had made a sweetheart deal with the City of New York to get the franchise without bidding. This deal was so bad for Sterling that it almost guaranteed that the company would continue to lose money. Besides that, the City of New York found out about Time Inc.'s plan to sell its cable TV franchise, Sterling Manhattan and Home Box Office, and refused to approve of it.

Thus, Time Inc. was forced to hang on to Sterling Manhattan Cable and Home Box Office. Some people at Time Inc. thought it had been left holding the proverbial bag. This is why I refer to Home Box Office as the almost accidental company, because, aside from facing stiff competition in the marketplace, it was beleagured by its competitors, overregulated by the government, and unloved by its own parent, Time Inc.

Still, not everything was bad for pay cable in 1973. A little more than a week before HBO came to life, another launch was announced that would change the cable industry forever. The announcement was made by the White House, the FCC, and the National Aeronautics and Space Administration (NASA), and the message was short but sweet: NASA was going to launch space satellites. The FCC said that these satellites would link cable systems together to create a new communications and entertainment network. NASA's spokesman, Dave Silverman, said the first satellite, or "bird," would be sent aloft in April of 1974.

That bird would become a golden goose for HBO.

Chapter 5

Going Up on the Bird

It was in 1974 that the future of pay cable TV was decided. It was the year that would launch HBO into leadership of the industry, a preeminence that HBO has not given up. It was the year that would establish Jerry Levin as a legendary figure in the history of cable television.

Cable TV had been at war with its competitors for a long time, but it was, slowly but surely, making progress. Its competitors were becoming more and more panicky. By 1974, there were 67 million American homes with commercial television, and it was estimated that a little under 10 million also had cable. Of those homes with cable, only 190,000—or less than 2 percent—had premium pay cable. Clearly, HBO shouldn't have seemed like a big threat to its competitors, but it had them running scared.

If they had known what was coming, cable's competitors would have been *sprinting* scared. The milestone for cable and for broadcast television in 1974 was the fulfillment of a vision first proposed by Arthur C. Clarke: a space satellite hanging in the sky 22,300 miles over the equator. This satellite continues to act as an electronic mirror, bouncing TV signals received from a single point on the earth back to many places on the earth. (The signal is like a rifle shot going up and a shotgun shot coming down.) Each space satellite has many of these electronic mirrors (or transponders) on it, usually somewhere between twelve and forty-eight of them. Now governments or private companies, such as Western Union and RCA, can pay some space agency, such as the National Aeronautics and Space Administration (NASA), several million dollars to put a satellite into orbit.

Then the transponders or electronic mirrors are leased to customers who want to use them to transmit telephone calls, computer data, radio signals, or television pictures. The signal goes up to the transponder on the satellite from one master transmitting earth dish and is reflected down to scores or even hundreds of receiving dishes scattered all around the country.

The first communications satellite, or "bird," as it is called in the industry, to have an impact on Jerry Levin was a Canadian one. It was used to transmit a 1973 speech by House Speaker Carl Albert in Washington, D.C., to the audience at the National Cable Television Association meeting in Anaheim, California. Levin watched the speech and was stunned by the clarity of the picture that had traveled 22,300 miles up into space and 22,300 miles back again.

By early 1974, Western Union had launched the first U.S. domestic communications satellite, Westar I, and RCA announced it would follow suit within the year. Thus the hardware was in place; now all that it lacked was somebody who could make economic sense of it. That somebody was Jerry Levin, who wanted to use a satellite to turn HBO into a national entertainment communications network.

But first Levin had to sell the directors of Time Inc. on making the kind of financial commitment necessary. The Time Inc. executives to whom Levin made his presentation admitted later that they were fascinated by what Levin had to say but didn't understand much of it. This was one of Levin's great skills: He could communicate a concept with such passion and commitment that, even if the details weren't understood, one had to believe that he was right.

Time Inc.'s directors told Levin to go ahead if he could show them that the scheme would pay off financially. After a period of number-crunching, Levin and his staff eventually projected that going up on the satellite would pay off. Again the Board wasn't quite sure what it was approving, but in a tribute to Levin's persuasiveness and the Board's courage, it gave the project the go-ahead.

The two main costs were getting up on the bird and getting down off the bird. Basically, it took one transmitting dish to get up to the transponder; getting down was not quite so easy. That required having a receiving dish at every cable system location—lots and lots of receiving dishes.

The cost of getting up on the satellite was going to be high. As already mentioned, Western Union was the first company with a satellite actually in orbit: Westar I, with twelve channels. The competition for those channels was stiff, and that meant that Western Union could ask a lot of money for leasing a channel. Fortunately for Levin and HBO, the RCA satellite due to be launched December 1974—called Satcom I—was going to have twenty-four channels on it.

RCA had $150 million riding on Satcom I and it was eager to find customers. Enter Jerry Levin. By the time RCA and HBO had completed their deal, HBO had agreed to pay $7.5 million for one channel for six years. That's $1.25 million a year, which is a lot of money for a company grossing $300,000 a month. The rental of the satellite channel was going to cost HBO five months' income.

The second hurdle was getting the cable operators to buy the receiving dishes to pull the HBO programs off the satellite. The choke point on this aspect of the plan was the Federal Communications Commission. The FCC insisted that the receiving dishes must be at least twenty-five feet in diameter. The price tag on these was $100,000 each, which was too expensive for most cable operators. Levin persuaded a leading dish manufacturer, Scientific-Atlanta, to cut its price significantly for a bulk order.

Meanwhile, Jim Heyworth, one of Levin's team, made a survey of some Florida cable operators who had been unable to receive HBO. They agreed to take the service if HBO would deliver its signal by satellite. Heyworth's financial analysis concluded that the revenues from the Florida cable systems would justify going up on the bird.

The reaction of other cable operators was also favorable. UA-Columbia said it would install receiving dishes for its cable systems in Florida, Arkansas, Texas, Arizona, California, and Washington State. TelePrompter joined in and gave Scientific-Atlanta an order for fourteen receiving dishes—its biggest order to date.

As HBO prepared to go up on the bird, two other matters consumed Levin's time: programs and permissions.

HBO still needed the permission of the FCC to transmit its programs via satellite, but the situation was becoming bizarre. On the one hand, HBO was asking permission to go up on satellite, and on the other hand, HBO had a lawsuit pending against the FCC challenging the federal agency's right to regulate pay TV programming at all. Further, after HBO asked permission, the FCC notified the whole television industry that it had the chance to file objections to HBO's plans if it wanted to do so. Given the vehement network and theater owner opposition to pay TV, one would have expected a firestorm of protest. Instead there was absolute silence. Levin was incredulous.

Most of those opposed to pay television, such as the movie studios, theater owners, and broadcast networks, clearly didn't understand anything about the new technology and its implications. Or, if they did understand it, they didn't care, because they didn't think Home Box Office could make it work financially. Earlier I wrote that one of the three factors in the amazing success of Home Box Office was the stupidity of its competitors. This mute response to the FCC's invitation to protest HBO's request to go up on satellite is a classic illustration of that stupidity.

For the program selected to showcase the new delivery system, Levin did another smart thing. He picked a single event that had already been hyped extensively all around the country and in all the national media. He made a deal with flamboyant sports promoter Don King for the rights to the Muhammed Ali–Joe Frazier fight set for the Philippines—the Thrilla from Manila.

The choice was inspired because it clearly demonstrated the virtue of satellite broadcasting. Here HBO was showing something *live from overseas,* and none of the networks could do it. Because they weren't up on the satellite, the networks would not be able to show anything about the fight until a film of it was flown across the Pacific and physically handed to them. HBO, in contrast, would show the fight live to its subscribers as it happened. HBO was going to scoop all three broadcast networks on the biggest news story of the month, and it wasn't even in the news business.

There was a big party set for 9:00 P.M. on September 30, 1975, at the Holiday Inn at Vero Beach, Florida. Interested cable operators from all over were there as HBO's guests. The signal would beam up from the Philippines, bounce off the satellite, and beam down to receiver dishes to be fed into the cable systems in Fort Pierce, Florida, and Vero Beach.

The event began with speeches from Time Inc.'s top executive, Andrew Heiskell, and the chairman of the FCC, Richard Wiley. That was followed by two movies, *Brother of the Wind* and *Alice Doesn't Live Here Anymore.* Then, some more speeches and, finally, the Thrilla from Manila.

When the fight started, everybody had the same reaction that Jerry Levin had had almost two years before as he sat in Anaheim,

26 *Inside HBO*

California, watching House Speaker Carl Albert talk to the cable association convention from Washington, D.C. The viewers at Vero Beach were amazed at the perfect picture they saw from thousands of miles across the Pacific Ocean.

For Jerry Levin, going up on the bird was one of the great business decisions of the decade. Paul Kagan, industry analyst, characterized the significance of what Levin and HBO did as being a simple business decision that changed an entire industry.

The amazing thing about HBO's going up on satellite was that it took its competitors so long to follow suit. The most aggressive was Ted Turner, who didn't get his superstation onto satellite until over a year later, in December of 1976. HBO's direct competitor, Viacom's Showtime pay movie channel, didn't go up on the bird until March of 1978.

HBO's going up on the bird was made more courageous as a risk because HBO still wasn't making a profit. It had lost $1 million in its first full year, 1973; $4 million the following year; and $3,150,000 in 1975.

Still, there were some figures for Levin to love. As a result of going up on the bird, new HBO subscriptions rocketed to an average of 30,000 a month and HBO ended 1975 with 287,199 subscribers—500 percent growth in just one year. But it was nothing compared to what was still to come.

Chapter 6

The Rush to the Sky

The success of the Thrilla from Manila and the sky link by satellite brought jubilation to the Time-Life Building in New York City. It also triggered something explosive in the mind of a southern maverick named Ted Turner.

Ted Turner is many things to many people: championship yachtsman, egomaniac, playboy moralist, baseball club tycoon, cable TV maven, right-wing saint, genius, hustler, salesman, Don Quixote, and The Mouth of the South.

To everyone, however, Turner is a man with a knack for business. Turner inherited a small outdoor advertising business in Atlanta from his father and turned it into a financial success with a lot of hustle. Ted did so well that he was able to buy a down-at-the-heels UHF television station in Atlanta, WTCG. The station ranked fifth in a city with five TV stations and moved to fourth only when one of the other stations folded. If WTCG, which he renamed WTBS, was a loser, its new owner was not. Whatever else Turner is, he is not willing to admit there is anything he cannot do if he wants to do it. He is a living example of the American entrepreneur—unfettered by committees, stockholders, and corporate naysayers.

In 1975, Turner's visionary button got pushed by what Jerry Levin did with the bird in the sky. WTBS was already sending its signal by microwave to nearby cable operators. When he heard about HBO using the satellite, Turner immediately realized that he could turn WTBS into a national station by feeding to cable systems all over America. His idea was to beam the regular programming of WTBS—mostly old movies and situation-comedy reruns—plus the games of the two teams he owned, the Atlanta Braves and the Hawks.

27

He would charge the cable operator a small fee and the cable subscriber nothing. Most of his profit would come from selling ads to national advertisers. He could become a national cable network all by himself and charge advertisers only a fraction of what the bigger broadcast networks did.

Turner hated the networks and was constantly accusing them of undermining American values with the sex and violence of their programs. This made him the bane of the network executives and the darling of the political conservatives.

And, back in 1975, it was an uncomfortable time for his critics in broadcast television when Turner announced he wanted to turn his WTBS into a national superstation.

Maybe because they learned from what HBO had done, the networks fought Turner every step of the way. They started out not liking this cornpone from the South and ended up being frightened of him. Unfortunately for the New York–based networks, the era of the satellite had dawned. Not only couldn't the networks hold it back, they would soon have to join in the rush to the sky.

By the end of 1976, Turner got his WTBS up on satellite and dubbed it the first "superstation" in the nation. It would turn out to be enormously profitable, and WOR in New York and WGN in Chicago would follow his example. Soon after that, Turner would stun the television world by launching the first and only twenty-four-hour TV news channel, Cable News Network (CNN).

Critics said the public wouldn't support a twenty-four-hour TV news channel, but once again the critics were wrong, so wrong that ABC would try to go into competition with CNN only to go down in flames and NBC would follow only to be outfoxed by the good ole boy from Atlanta.

Meanwhile back at HBO, things were looking better and better. The number of new subscribers had risen to thirty-six thousand a month, and nine months after the Thrilla from Manila kickoff, HBO had grown to five hundred thousand subscribers.

The one thing that was holding HBO back from exploding in growth was still the high cost of receiving dishes for the many cable systems that now wanted HBO. Working with the National Cable TV Association (NCTA), HBO got the FCC to relax its technical requirements enough to allow a smaller and less expensive receiving dish. Finally, in time for Christmas of 1976, the FCC approved a dish of half the size—and less than half the expense.

Aside from the technical problems, HBO continued to struggle with the programming problem of getting enough to satisfy its subscribers. The mix that HBO subscribers were offered was eclectic to say the least. For instance, in 1976, HBO subscribers saw thirteen American Film Theatre plays; the BBC series "The Pallisers"; twelve Playboy Club comedians; a Bette Midler concert; the Folies Bergère from Las Vegas, and fourteen showings of *Gone with the Wind.* Of course, in 1976 HBO subscribers also got to see Cowtown Rodeos from Woodstown, New Jersey; local roller derbies; and many second-rate movies repeated too many times to count.

Something else happened at HBO during this time. HBO people began to take on airs and become aggressive, abrasive, and arrogant. The attitude was perhaps understandable—after all, they had struggled long and hard and they had made it. Still, it wasn't pleasant, and it would hurt HBO and Time Inc. in the years to come.

In 1976, Munro moved Jerry Levin upstairs to be Chairman and CEO of HBO and put N. J. "Nick" Nicholas, thirty-six, in as president and chief operating officer. Nicholas came from several financial jobs in the Time Inc. organization, with the last being that of president of Manhattan Cable. He fired almost all the HBO executives he found upon arrival and brought in his own team. Most notably, he brought in Tony Cox, thirty-four, from *People* magazine to work on HBO affiliate relations; Michael Fuchs, thirty, to be director of special programming; and Austin Furst, thirty-two, as vice president of programming. Nicholas said it was Furst who turned HBO around after a disastrous 1976.

During this time down in Washington, D.C., the broadcasters, theater owners, and cable people continued fighting one another.

The Congress, the Justice Department, and the Gerald Ford White House sided with the cable people, but the FCC tended to side with the broadcasters. The Congress and the Justice Department felt it would add competition to the stultified television entertainment market, and the White House hated the pro-liberal Democratic bias that it perceived among the broadcasters. On the other hand, the FCC seemed to have become the chief promoter and cheerleader for the industry it was supposed to regulate. (There is nothing unusual about that; it has happened with many regulatory agencies.)

The matter of cable regulation was ultimately not solved by the FCC but by an HBO lawsuit *against* the FCC. It took a lot of money and time, but HBO finally won the suit and the FCC essentially deregulated much of cable.

Aside from the legal and government regulation battles, cable was doing better than expected in the late 1970s. *Business Week* reported in August of 1977 that cable had grown to a $770 million business, making huge profits.

By the end of 1977, one out of every seven homes was connected to cable. HBO had reached the level of six hundred thousand subscribers. It began with about one-twentieth of 1 percent of that many customers five years before, and it would have twenty times that many customers five years later. That is unprecedented growth in five years, especially for an almost accidental company.

Chapter 7

Profits for HBO

One of the factors of Home Box Office's extraordinary success was the willingness to take risks displayed by its people from time to time. It was obviously a risk to commit to going up on the satellite, and that paid off. Another risky move that proved to be a good one was suing the FCC.

Home Box Office challenged the right of the FCC to protect the broadcast television industry to the detriment of the cable television industry. On March 29, 1977, the U.S. Court of Appeals agreed with Home Box Office and struck down FCC rules that protected the commercial TV broadcasters from competition. This ruling became known as the HBO Decision. As soon as it was handed down, the broadcasters suddenly knew that it was time to get into the cable business.

An even more important milestone was passed in October 1977, when HBO turned a profit. Echoing *Sometimes a Great Notion*, the title of the first film shown on HBO, it was now a profit center. And that's what mattered to Time Inc. Even better, HBO quickly became an incredible cash cow, generating bushel baskets of profit for Time Inc. and eventually outstripping the all-important magazine division.

To understand the importance of the cash flow generated for Time Inc. one has to understand that many businesses operate on credit and often have to wait a long time to get paid. This means that these businesses must go to banks to borrow money against what is owed them and must pay substantial interest on this borrowed money, which cuts into their profits. With HBO, a bundle of cash came flowing into Time Inc. every month, a state of affairs that was very pleasing to the fiscal conservatives who ran the corporation.

32 *Inside HBO*

The year HBO turned a profit, Tony Cox was brought over from *People* magazine. Tony's job was to hype Home Box Office to cable operators and get them, in turn, to hype HBO to their subscribers. Tony was good at his job, although some said he could be a bit heavy-handed at times, using Time's clout to keep some recalcitrant operators in line.

Tony's job at HBO was not unlike what he had done at *People*. In the simplest terms, one reason *People* is a success is that it is so easy to buy it. Tony would sell supermarkets on letting his *People* magazine rack be at eye level by the checkout counter. Grocery customers waiting in line to check out found themselves face-to-cover with *People*. The provocative pictures and titles did the rest.

What people can't see or find, they can't buy, and what they do see, they may buy. At HBO, Tony's job was to make sure that potential subscribers knew about HBO and how easy it was to connect up to it.

About the time Tony was brought over from *People* there was another significant personnel change. Michael Fuchs was hired from the William Morris agency to help with talent and programming (after his boss at William Morris turned down the same job). Fuchs is probably the most controversial executive at Home Box Office, if not in the entire cable industry. Intensely competitive and devoted to Home Box Office, he has a confrontational style of negotiating that has made him unpopular in Hollywood and the terror of some of his own colleagues. Yet it would be unfair not to note that he has had much to do with the success of HBO in the ten years since he began dealing with programming there.

Some of the "old" faces moved on. Nick Nicholas, having managed to put HBO in the black, was moved up to be its chairman of the board. Nicholas was getting a reputation at Time Inc. as a solid organizer and manager with the ability to focus on the bottom line successfully. This reputation would stand him in very good stead in the decade that followed. Jerry Levin, the Father of Pay Television II, was moved from chairman of HBO to become head of the Time Inc. Video Group. From there he oversaw both HBO and the Time Inc. cable systems company, American Television and Communications (ATC), which is the second largest cable system conglomerate in the country. (Time Inc. had sold its 9 percent of ATC in 1973 and, four years later, bought the entire company.) And Jim Heyworth was moved into the presidency of HBO.

By 1979, HBO had signed up its one thousandth affiliate cable system, owned by Polly Dunn of Columbus, Mississippi. By the next year, HBO had 4 million subscribers and was on cable systems in every one of the fifty states. Home Box Office had indeed come a long way in eight years.

Of course, there was continuing concern over churning—a lot of unhappy subscribers continued to disconnect. But the new ones were signing up faster than the old ones were zapping their subscriptions. As long as the industry could keep spreading into new territory, HBO could add new subscribers. That was a fact that would not dawn on HBO executives and other cable programmers until four years later.

After the courts sent down the Home Box Office Decision, all three networks and many other major corporations wanted to get into cable. NBC and CBS said they would soon try to bring culture to America with opera, ballets, and subtitled foreign movies. CBS started a cable unit in the spring of 1980 and asked the FCC for permission to get into the cable business.

ABC, who understood that America loves car chases and semi-violent sporting events, established its ABC Video Enterprises in 1979 and bought into an all-sports cable service, the Entertainment and Sports Programming Network (ESPN). ESPN was owned mostly by Getty Oil, which was interested in getting into cable and had bought into ESPN for less than $10 million. Getty had in mind a merger of its oil money with the creative talents of several movie studios into a company that would blast HBO out of the marketplace. They would call this enterprise "Premiere."

Also going on at the Time-Life Building in the spring of 1980 was the beginning of another cable network, which was to be known as "BBC in America." It would consist of Time-Life Films distributing BBC programs to cable. But "BBC in America" died, mostly because there was confusion about corporate objectives within Time Inc. While one part of Time Inc. was trying to sell Time-Life Films because Time Inc. didn't want to be in the movie business, another part was trying to get into the movie business with the BBC. The BBC would eventually find a home across the street from the Time-Life Building. BBC made a deal to provide all of its programs exclusively for ten years to a new cable network, Radio City TV (RCTV), headed by ex–CBS president Arthur R. Taylor. RCTV later changed its name to The Entertainment Channel.

34 Inside HBO

In the cable industry at large, everyone was optimistic as the National Cable Television Association (NCTA) convention met in May 1980 in Dallas. One sign of the boom mentality was the presence, for the first time, of a lot of bankers looking for places in cable to loan money. *The New York Times* made it official when it ran a story about how the new TV technology was changing the way movies are delivered to the consumer. This change, in turn, was causing a shift in who was going to be powerful and rich in the entertainment industry.

At Home Box Office, it was decided to try creating a second pay cable TV service. HBO had tried that once before with Take Two, but it failed and was dropped. The second try, called Cinemax, was started in mid-1980. It was envisioned as an all-movie service that ran twenty-four hours a day offering classic and "ever-popular" movies. Translated into English, that means running old movies. HBO's main competitor, Showtime, countered by announcing it would go to twenty-four hours a day starting in January 1981.

Other companies were merging to strengthen their cable market position. In late 1981, Westinghouse Broadcasting and Tele-Prompter got together to make the biggest merger in cable history up to that time. Westinghouse paid $600 million for the giant cable company in which Jack Kent Cooke, the sports millionaire, made his fortune—or at léast one of them. (It enabled him to buy the Washington Redskins football team and split up with his longtime wife in the biggest divorce settlement in the history of mankind, according to the *Guinness Book of World Records*.)

In another joint deal, ABC and Westinghouse tried to out-Turner Ted Turner. The two broadcasting giants got together in late 1981 to form a twenty-four-hour, all-news TV network called Satellite News Service. And *Playboy* magazine and Rainbow Programming said they would be getting together to transform the Escapade Cable Network into a video cable version of *Playboy*. Rainbow was run by none other than the man who dreamed up the idea of HBO: Chuck Dolan.

By January 1981, the total number of cable subscribers in the United States was up to 17 million, or 50 percent more than there had been five years before.

Time Inc. announced that two more major video projects were under development. One was a teletext system to be tested in San Diego that could transmit up to five thousand pages of information to subscribers. The other, announced June 25, 1981, was a cable TV magazine called *TV-Cable Week*. In addition, in August 1981, Time Inc. shelled out $15 million to Columbia Cablevision for 50 percent interest in the USA Cable Network and turned around in October to sell a third of it to MCA.

Out in Hollywood, the studios had joined together to sue the Sony company for making videocassette recorders (VCRs) that allowed people to tape programs in their own homes. The studios claimed that VCRs were an infringement of the right of those who owned the copyright to those programs—frequently the Hollywood studios themselves.

So, there was happiness in Hollywood in October of 1981 when the United States Court of Appeals ruled that the videotaping of copyrighted TV programs was illegal. Hurrah, hurrah, said the studios. Less than eighteen months later the studios would be grateful that videocassette players existed.

PART II

THE GOLDEN YEARS: 1979–83

Chapter 8

Cable TV—Promise or Phantom?

Cable has been an industry filled with promises and phantoms. The biggest problem is trying to figure out which is which.

The cable business has been plagued with problems, bad news, negative images, and lousy public relations, yet somehow it has survived. Viewers still seem endlessly fascinated by it.

Thomas Bedell in *Staten Island* magazine explained the appeal of cable television in five words: "More sports. More movies, too." Rod Warner at Storer Cable put his finger on one of the strange but true conditions of the television world when he said that commercial TV is so bad, people are hungering for an alternative, and they turn to cable in the hope that they will find it. Unfortunately, the public's enthusiasm for cable has often been fired by what they hoped to find rather than what they actually found. He said viewers just wanted something new and better, and they didn't really know the difference between WTBS, ESPN, or IUD.

The sad part of this theory is that cable television in 1981 was not prepared or even inclined to offer America anything that was better than commercial television. Jerry Levin had envisioned cable as a chance for television to be creative, novel, and stimulating, but it appeared to be offering viewers more of the same old thing.

In spite of this, business was good for cable TV in the late 1970s and early 1980s. The FCC reported that, in 1979, the cable industry had grown 20 percent over the year before and was doing $1.8 billion in business and had assets of $3 billion. The National Cable Television Association reported that cable subscriptions had grown from 14.1 million homes at the end of 1979 to 17.7 at the end of 1980, with the figure projected to be 20.6 by the end of 1981.

This kind of growth was the result of cable operators' getting franchises to build exclusive cable systems in hundreds of cities and towns around the country. To get these franchises, cable operators were competing vigorously, and each was outpromising the others about all the bells and whistles they would provide. Unfortunately, too often the promises were out of synch with the economic reality of running a profitable cable system.

To the outside observer, a cable TV franchise is a license to steal. Donald Nelson, an Indiana legislator, commented in December 1980, "A cable television franchise is like a liquor license—very lucrative." And the mayor of St. Louis (which still doesn't have cable TV) said, "This is the last gold rush—or the latest one, anyway."

A franchise hustler for a cable company wrote a magazine article in May 1981, "Dirty Tricks: The Confessions of a Cable Franchise Negotiator," under a phony name. In the article he spelled out what was going on in those frenetic days. He said he doubted if half the promises made to cities by cable operators would ever be kept. They would go bankrupt if they did, he said. He concluded:

> It's just like politics. Everybody makes extravagant promises, but as soon as they're elected they start making excuses why the promises can't be fulfilled.

To put this in perspective, however, one should keep in mind that the local politicians are also to blame for the broken promises. When the politicians in one town would hear that a nearby town got a cable operator to promise a system with thirty-six channels, the politicians in the second town would demand forty-eight. Within some cities or counties, individual rivalries would have the school board, fire department, police department, agricultural extension service, and so on each demanding a separate channel of its own to compete with the city council and the board of supervisors, which each had a separate channel.

After a while, some major cable operators such as American Television and Communications, a subsidiary of Time Inc. and sister company to Home Box Office, dropped out of the competition for many cities because the demands of local officials were just too ridiculous.

Baltimore is an example of the franchising process run amok. I attended the preliminary hearings that the city held for anyone interested in bidding on its cable franchise, and seventy-three different potential bidders attended. Weeks later, after the city had gotten through listing all of its demands—including cash bonds and advance cash deposits—it ended up with only two bidders. One of those was a local group hastily put together and laughingly referred to by knowledgeable observers as three guys and a clipboard. This scenario is typical of many misguided cities.

Even with all of HBO's problems, however, in 1981 everybody wanted to be in cable TV. The last network holdout, NBC, announced in May of that year that it would become a full partner with the Radio City TV programming service started by Rockefeller Center the previous December and named The Entertainment Channel (TEC).

While it may have looked as if broadcasters were jumping into cable in a mad rush in 1981, the rush actually had been going on for a long time. The March 23, 1981, issue of *Cablevision* magazine reported that broadcasters owned 40 percent of the thirty-six hundred cable systems then in the country and that broadcasters, in spite of their public badmouthing of cable TV, had been major owners of cable systems since as far back as 1967.

As for predicting the future of cable TV, David S. Abbey, senior VP with the Katz Advertising Agency, made a prediction in 1981 that nobody seemed to pay much attention to at the time. He said:

> Pay television may fade as consumers choose to buy programs on videocassettes or videodiscs rather than as a wire service.

But in 1981 the industry ignored this prediction. As little as three years later they would wish they hadn't.

Chapter 9

The Infant Becomes a Giant

It was 1981, and Home Box Office was on the crest. It was the most successful pay television enterprise in the history of the world and a major profit center for Time Inc. It was feared and hated in Hollywood.

In fact, some feared (and others hoped) that this nine-year-old pay TV service would *take over* Hollywood.

Some observers have divided the life cycle of HBO into three five-year segments: 1973–1977[1], trying to stay alive; 1978–1982, hauling in big profits; and 1983–1987, fighting off competition and diversifying beyond cable systems.

Success did not always come easily to HBO, especially with all their competitors. One kind of competition came from the cable operator himself. The thinking was: If you owned a lot of cable systems, why not make your own deals to rent movies from Hollywood instead of going through a middleman like HBO? The Times-Mirror Company owned a number of cable systems, and it bought into a programming service called Spotlight for just that reason.

It began to use Spotlight to replace existing pay services on its cable systems, and often the service replaced was HBO. The point of the experiment was to see if people would be happy with a movie cable service even though it was not the highly touted HBO.

---

[1] 1973 was HBO's first full calendar year of existence. It actually began on November 8, 1972.

The Infant Becomes a Giant   43

To combat this kind of competition, HBO advertised "exclusivity" and demanded exclusive cable TV rights from the Hollywood studios for the movies it was renting.

Another approach used by HBO to dominate the market was to finance movies itself. By doing so they could also beef up their supply of programs and movies, of which there never seemed to be enough to fill the demands of cable TV. But original productions cost a lot of money if they are going to be good enough to attract customers. Original programs were just a drop in the bucket. To fill the hours HBO continued to depend on repeats.

Jack Curry, a reporter for the New York *Daily News,* observed in a 1981 story, "Show me a Home Box Office patron, and I'll show you someone who has seen *The Great Santini* 15 times."

Curry said a mouthful. If a subscriber were to be enough of a movie nut to subscribe to all three major pay channels—Showtime, The Movie Channel, and Home Box Office—during the six months ending in March of 1981, he would have had the chance to see:

Rocky II	44 times
Superman	53 times
The Muppet Movie	37 times
Close Encounters of the Third Kind	41 times
. . . And Justice for All	39 times
The Rose	38 times

Or, six movies for a total of 252 times. Whew!

A classic example of what a shortage of programming can do to a pay cable service came on May 11, 1981, as HBO gave its viewers another top-flight boxing match, for which it takes pride.

It was the Gerry Cooney–Ken Norton fight from Madison Square Garden. HBO had paid $550,000 for the cable rights, but they considered it a sound investment. After all, the Home Box Office Sports Department figured they would be able to fill the hours between 8:30 and 10:30 P.M. Unfortunately, Cooney took just fifty-four seconds in the first round to finish off Norton, which left HBO and its three announcers with two hours to fill. They had no alternative programming ready.

That night viewers who stuck with HBO saw numerous replays of the fight, heard speculation after speculation on Cooney's and Norton's past, future, and present, and listened to limitless pontification over the nature, direction, history, and future of the fight game.

By August of 1981, HBO was negotiating with 20th Century–Fox Studios to start on a new project, namely, movies made specifically for pay TV.

Another irony in a parade of ironies was that the man with whom HBO was negotiating at Fox was the new head of the studio, Alan J. Hirschfield. Back in 1978, Hirschfield had been at Columbia Studios when the David Begelman scandal broke. Columbia executive Begelman had been embezzling money on a grand scale, and Hirschfield wanted to fire him. (Actor Cliff Robertson, who is very straight about accounting for his money, blew the whistle on Begelman and hasn't acted in a major Hollywood movie since.) However, the studio owners wanted to fire Hirschfield for wanting to fire Begelman. So, Hirschfield went to Jerry Levin and Dick Munro at HBO to get them to buy Columbia Pictures. It didn't work out, and Hirschfield left Columbia to go to 20th Century–Fox, and here he was negotiating with HBO again.

Yet another irony is that Time Inc. had a potential source of programs in its own in-house movie operation, Time-Life Films, but it had gotten rid of it. And way back in 1967, Time Inc. had owned 6 percent of MGM as part of a plan with Edgar Bronfman, the head of Seagrams, to buy MGM and merge it with Time Inc. Instead they lost out to a Las Vegas investor, Kirk Kerkorian.

Time Inc. made another boneheaded move when it virtually gave the rights to the Time-Life film library to Austin Furst when he left to set up his own home video company, Vestron. It sold the rest of the operation to 20th Century–Fox. One of the rationales for this sale was a signal to the movie industry that HBO would stay out of the movie production business if the movie studios would stay out of pay cable. The message didn't stick well on either side.

In 1981, insiders knew that HBO wanted its own movie studio. As one of them put it, HBO needed a production company that could stick to budgets and was willing to tackle subjects that the commercial networks were afraid to handle.

Early in the year, Time Inc. reorganized its Video Group as part of its plan "to become a broad-gauged entertainment and information programming company." Jerry Levin, the Video Group's vice president, said that a new development staff was being formed to look into other businesses that were related to television. This included foreign cable, home video, and even theatrical exhibition.

One of the big deals under consideration during the summer was one that had the industry buzzing: the possible buy-out of the USA Network by Time Inc. USA Network had been started in 1977 by United Artists, Columbia Pictures, and Madison Square Garden. At the time, Columbia was in the middle of an internal furor over the David Begelman embezzlement. The other partner, Madison Square Garden, was in financial trouble because its teams had not been doing well.

Meanwhile, as talk of Time Inc.'s buying USA Network was going on, the first of a recurrent speculation surfaced on Wall Street, as reported in *Forbes* magazine. Namely, that Time Inc. itself might be a juicy target for someone else to buy.

Forbes made the point that Time Inc., which everyone in the general public thinks of as basically a magazine company, was now actually a cable TV company first; a timber company second; a paper milling and cardboard company third; a magazine company fourth; and a book publishing company fifth.

Forbes said that Time Inc. dismissed takeover talk with the answer that the company's assets go home every night—referring to its magazine editors, reporters, writers, etc. However, *Forbes* noted that the magazine wasn't king there anymore and that television cables and trees in Texas *didn't* go home every night.

What's more, said *Forbes,* "TV types are less sentimental than writers and editors."

Chapter 10

HBO and Hollywood

If ever in the history of American business there were two segments of the same industry that hated and feared each other more than HBO and Hollywood, it has escaped general attention.

HBO and Hollywood had the symbiotic relationship of mutual parasites who loathed each other but couldn't survive without each other. To use an old Texas expression, they were like two strange dogs on their first meeting, circling and sniffing, sniffing and circling. Each one had no greater pleasure than lifting his leg on the other.

According to the investment firm of Goldman, Sachs and Company, in 1981, the balance among the Hollywood movie industry, the theater owners, the cable industry, and home video industry was shifting. It said that several years before, there were only a million homes with pay TV and predicted growth to 27 million by 1985 and 46 million by 1989. These pay TV subscribers would become used to uncut, commercial-free movies coupled with the convenience of seeing them at home. Result: Movies in theaters and on commercial TV would become less desirable, and that eventually, in turn, would affect the profitability of commercial TV and theaters.

The studios were becoming worried about the money power of HBO and other cable programmers. They saw the number of seats in theaters dwindle while the number of people connected to cable TV grew. Some studios saw that they had to get into cable TV to combat the power of HBO.

What happened might be called "The Movie Empire Strikes Back." The movie moguls tried to form their own pay cable TV service, which they called Premiere. The players were 20th Century–Fox, Universal, Paramount, and Columbia, all of whom joined forces with the Getty Oil empire to challenge the pay cable supremacy of HBO. The plan stunned the industry, particularly HBO. If four of Hollywood's biggest seven studios had been allowed to combine with the resources of Getty Oil, it could have seriously hurt HBO.

Fortunately for HBO, Uncle Sam stepped in. The new company was challenged by the Justice Department as a possible restraint of trade, and hearings were held before New York Federal District Judge Gerard Goettel. Eight and a half months later, on New Year's Eve of 1980, Goettel ruled that the Premiere venture was a violation of the Sherman Anti-Trust Act.

One reason for Goettel's ruling was that the studio partners in Premiere agreed not to sell films to any competing pay cable services for at least nine months after the films were aired on Premiere. Just to get some idea of the kind of films we're talking about, the year before, eight of the ten most successful films (and all of the top five) came out of these studios. Films such as *The Empire Strikes Back, Kramer vs. Kramer, The Jerk, Airplane,* and *Smokey and the Bandit.*

The executives at Premiere claimed that they needed the nine-month "window" so that their service would be offering something different enough from their competition and attractive enough for potential subscribers that Premiere would have a chance to succeed against established services, primarily Home Box Office.

The judge admitted that Home Box Office had almost 70 percent of the pay cable subscribers, but he didn't see that as something in restraint of commerce. His reasoning was that even though HBO had almost 70 percent of the current subscribers, the potential subscribers still had a lot of services to pick from.

Goettel believed that Premiere would torpedo both Home Box Office and Showtime. "If Premiere were allowed to continue operating . . . Showtime and The Movie Channel could be put out of business . . . [and HBO's] subscribers will undoubtedly begin to disconnect."

The president of Premiere, Chris Derick, reacted to Judge Goettel's ruling this way: "The ruling hands over the pay TV business to Time Inc."

If that was true, Goettel suggested that Time Inc. had earned its position in the pay cable business. In his written opinion, he said:

> The popularity of [pay TV] has grown so rapidly that it is not impossible that, by the end of the century, it will be the prime method for viewing motion pictures. This case is about who will reap the enormous revenues available from this enterprise.

Goettel went on to rhapsodize about how HBO had dared to start from scratch after others had failed; its courage in moving to satellite distribution; and all the rest of its pioneering skill. Having gambled and won, HBO should be where it was, suggested Goettel.

In a weird aftermath, the week after the judge's decision was handed down, a violinist dressed as a gypsy strolled through the California corporate offices of Premiere and played while the employees cleaned out their desks.

Late in February 1981, the Premiere partners filed a reply with the U.S. Court of Appeals, but a few weeks later the Court upheld judge Goettel's ruling. The entire Premiere project had cost the five partners involved about $20 million, including $5 million in legal fees.

Michael Fuchs, senior VP in charge of programming at HBO, told the Television Critics Association a few days after the court's decision that he thought the trial was a catharsis, that it had let people in the industry get their hostile feelings off their chests. That was good for everybody, he said, and besides, cable TV needed Hollywood, and Hollywood needed pay TV. The point missed by Fuchs was that while Hollywood did indeed need pay TV, it didn't need a colossus like HBO controlling it.

The four studios involved in the case had boycotted the pay services by holding back movies they hoped they could release first through Premiere. These films were now up for grabs, and HBO's Michael Fuchs wasted no time in talking to the four studios and dickering for those films; neither did Mike Weinblatt of Showtime.

Figuring that the studios were stuck with movies that needed to move out of storage and onto America's screens to recoup their losses, HBO and Showtime tried to drive very hard bargains.

One of the things on the side of the cable negotiators was that a movie is a perishable commodity. The older it gets, the less appealing it is to the public, until it gets very old. Then it may become a "classic" or "cult film" and regain value.

Besides that, HBO had been holding back its inventory of movies from its subscribers to make its supply of films stretch as long as possible in the case of a long court trial.

The Movie Channel had played the game a different way. The Movie Channel wasn't number one in the pay cable business, and it figured it had to try harder. Gambling on a quick court decision, it ran through its backlog of movies on its regular schedule. It was a way of attracting new customers fed up with the constant repeat plays on HBO and Showtime plus holding on to old subscribers.

As a result, after Premiere went belly-up, The Movie Channel was almost out of movies and ready to pay the studios a top price. While Fuchs at HBO and Weinblatt at Showtime were squeezing the studios in hopes of a bargain deal, The Movie Channel made a fast deal with Universal Pictures for seventeen of the movies it had been reserving for Premiere, including *Coal Miner's Daughter, The Electric Horseman,* and *The Blues Brothers.*

The Movie Channel (TMC) made a second deal a few days later for another seventeen-movie package from Paramount, another of the Premiere partners, for an estimated $5 million. The package was a particularly gratifying one for TMC because of the blockbuster movies it involved, including *Ordinary People, Urban Cowboy, The Elephant Man, Star Trek,* and *Airplane!*

Meanwhile, in the never-never land of this business, lengthy negotiations between Time Inc. and 20th Century–Fox had been going on: Fox was trying to buy Time-Life Films, which had lost $9 million the previous year. When that deal fell through, the negotiations shifted to Columbia Pictures, another Premiere partner.

In spite of the Premiere movie partners' boycott of the pay TV services for part of 1980 and 1981, Hollywood studios, including the Premiere four, doubled their income from selling movies to pay TV in 1980, something they had done every year since 1976.

Even with all this lucrative business going on, a lot of movie executives were angered by the outcome of the Premiere case and grimly determined to thwart HBO and Fuchs in whatever way they could. They began looking at various ways to maximize studio profits and to hurt HBO. This vendetta would not end soon.

For example, the people at 20th Century–Fox, where Fuchs was particularly hated, released the movie *9 to 5* for home videocassettes and videodisc just ten weeks after it was released to movie theaters. That was the earliest that any major movie had gone to the home video market.

This move was regarded by many observers as a clear signal to the pay cable mavens that, after the sabotage of Premiere, things were not going to return to business as usual. Boosting the home video business was a way to hurt pay cable and HBO, because who wants to wait for a good movie to show up on HBO when you can rent one from the video store six to ten months before it will appear on HBO?

By 1982, the pay cable TV world was changing. HBO was faced with increasingly tough competition.

Chapter 11

Original Programming

HBO started life as a program broker, but it didn't take too long for it to realize the necessity of creating programs of its own. For one thing, original programming was another step away from total dependence on the movie studios and sports as program sources as well as a step closer to the varied programming the commercial networks put on.

HBO has been doing original programming since 1975, with such programs as "Standing Room Only" and "On Location." HBO continued this into 1982, when it purchased eighty or ninety specials and some seventy original sporting events.

On the night of September 19, 1981, for example, half a million people jammed Central Park to hear Simon and Garfunkel in their first reunion in eleven years. HBO taped the concert for a February 21, 1982, release as "Standing Room Only: Simon and Garfunkel—The Concert in the Park."

Boxing continued to be a staple on HBO programming with both the Sugar Ray Leonard–Bruce Finch and the Roberto Duran–Wilfred Benitez fights scheduled for 1982. That's not as many fights as the nine scheduled in 1981, but that was more a function of the fight business than of the cable business. HBO was willing to schedule as many professional fights as it could arrange.

The value of boxing to HBO was demonstrated by the A.C. Nielsen survey done in October 1981 about the viewership of the Marvin Hagler–Mustafa Hansho and the Mike Weaver–James Tillis fights. Both fights outdrew competing commercial network shows.

HBO's share of TV households during those fights was 36

percent compared with 23 percent for ABC, 21 percent for NBC, and 15 percent for CBS.

Another study done by Beta Research, Inc., compared the audience for the Marvin Hagler–Vito Antuofermo fight with the Larry Holmes–Leon Spinks match on ABC the night before. Forty percent of the viewers preferred the HBO fight coverage, while only 19 percent preferred that of ABC. The remaining 41 percent weren't watching the fight at all.

Roger Director, writing in the February 13, 1982, issue of *TV Guide*, ranked HBO's sports programming—particularly boxing—high in comparison to other cable systems' *and* the commercial networks'.

Another effort in original HBO programming was the fictionalized nonfiction program. Sheila Nevins was in charge of HBO's special projects. One of the programs for which Nevins was responsible was "Flashback," in which historical documentary footage was mixed with dramatic re-creations to tell the story of some important past event, such as the crash of the Hindenburg or the fire at the Coconut Grove nightclub in Boston. While the shows used reality as a launching pad, they quickly shifted to dramatizations and animation to avoid boring the audience with facts. Nevins believed that reality had to be hyped up or "theatricalized."

Nevins's concept of "theatricalizing" news stories made the blood drain out of the faces of old-line, hard-truth journalists, but it had a legitimate precedent in the "nonfiction novels" (some term it "faction") of Truman Capote's *In Cold Blood* and Norman Mailer's *The Executioner's Song*. Both were based on actual news events, but were partially fictionalized.

Time Inc.'s second pay service, Cinemax, was created to compete with other pay services for people who wanted to take two pay-movie channels. It started out as a classic, oldies-but-goodies channel, and since it began—in August 1980—Cinemax has gained more than 2 million subscribers. Everyone agreed that Cinemax's growth was phenomenal, but the service has also had a high churn rate. In many cable systems it was around the 6 percent level per month.

Even so, many subscribers signed up for both HBO and Cinemax, which is exactly what the Time Inc. planners intended. They wanted Cinemax to be the number-two service to HBO, in place of Showtime or The Movie Channel or Spotlight. In fact, in order for Cinemax to achieve that position, some Time Inc. executives thought of releasing blockbuster movies on Cinemax first before releasing them on HBO.

This kind of thinking was part of the flexible marketing strategy that HBO followed to get the best results in a changing marketplace. Another example of this was when HBO began buying as many rights to as many movies and other programs as it could afford. The intention was to stockpile against the possibility of future shortages, labor unrest, or changing technologies, such as home video.

HBO tried to get all the exclusive rights to as many films as it could, but the studios didn't like selling all the exclusive rights to one customer; they felt they could make more money selling different rights to different customers. On the other hand, HBO was Hollywood's biggest single customer, so some compromises were made; the studios would sometimes give a little to keep HBO happy.

The 1983 annual national cable business survey, compiled by cable industry analyst Paul Kagan, showed how things were changing as HBO fought to keep their edge on competition. As HBO was turning ten years old, Kagan's survey of the twelve largest cable systems in the country revealed that 61 percent of new pay subscribers had joined either HBO or Cinemax. In almost all of them, HBO trailed behind Cinemax in new subscriber growth, but HBO was still far and away the biggest kid on the block.

At Time Inc. the Video Group (Home Box Office, Cinemax, and ATC cable systems, plus some miscellaneous video properties) accounted for $618 million of the total of $3.3 billion earnings of the company in 1981. That's 19 percent of the total gross earnings of Time Inc., compared with 14 percent the year before. The Video Group's share of Time Inc.'s operating *profit* was 42 percent, and that was up significantly from its 36 percent of 1980.

Review those last two paragraphs again. Most of us get turned off reading a lot of figures, but these underscore the significant status of television at Time Inc. While only 19 percent of the gross earnings of Time Inc. in 1981 came from television, 42 percent of the net profit came from television.

In other words, the Video Group at Time Inc. was enormously more profitable than was the Magazine Group. This fact grated on the stalwarts in the Magazine Group, many of whom had joined the magazine staff before some people at HBO had been born. The Magazine Group in general was envious of those gold-chained entertainment types that rode the elevator with them in the Time-Life Building and got off on the HBO floors.

The significance of the profit statistics was not lost on Time Inc. people or on those in the competing video and publishing groups. It was a trend that would have a severe effect on corporate decisions and would cost Time Inc. millions in corporate misjudgments in the months to come.

Chapter 12

The Money Side of Cable

HBO was a growing, constantly evolving young company, but it was not operating in a vacuum. The cable business was expanding as well. It was tempting outsiders and frustrating insiders. It was a business that was wild, weird, and wacky.

A clue to why many people were rushing into the cable business in recent years was supplied by the biggest cable system brokers in the country, Daniels and Associates of Denver. In a review of the 147 deals made in the five years between 1977 and 1981, Daniels revealed that a total of almost $1.5 billion was involved. Daniels reported that in 1977 cable company buyers were paying a price based on $391 for every subscriber. Five years later, that price had skyrocketed to $870 per subscriber, an increase of 123 percent every year! By 1987, cable systems in California and some other states were going for $1,400 per subscriber.

On the technological side, which is always of interest, ABC and Sony decided to make a technological leap that was experimental but could have had enormous implications for the future of home entertainment.

The videocassette recorder (VCR) was beginning to capture the interest of Americans, with about 3 million of them in homes around the country by 1982. By renting videocassettes or recording a TV program to be viewed later, the VCR allowed its owners to manage their TV viewing time. There was even a name for the new phenomenon—*time shifting*.

Still, there was the slight inconvenience with renting videocassettes in that you had to go to the home video store twice—once to pick it up and once to return it.

That's where ABC's Home View Network idea came in. It would deliver movies to your home electronically. During the night, while you were asleep, it would beam a signal directly to your home, turn your VCR on, and record a movie. ABC would make its money by scrambling the signal and renting a descrambler to the viewers.

An added appeal was that ABC expected to be able to get the rights from the movie studios to release films at the same time as they were released to the home videocassette market. This meant they could be in your home months before they would have been shown on HBO or on any other pay cable TV service.

The only problem with this way of delivering home video to the customer was that it didn't work. Labeling it TeleFirst, ABC and Sony gave it a test run in Chicago, and the public response was a giant yawn. After a few weeks' trial, the partners gave up on the project. No one is sure why it failed. It could have been too new, too complicated, or just too much too soon. (The concept would be tried again, as we'll soon see.)

As TeleFirst was failing, others were succeeding. The number of pay TV subscribers in the country had reached a record-breaking 17.5 million people.

Industry analyst Paul Kagan estimated that the average churn rate for pay TV at the time was 3.5 percent a month, indicating that 42 percent of those who signed up disconnected within a year. That meant that, in addition to the 7.1 million who signed up and stayed connected in 1981, about 5 million more had signed up but disconnected.

As the industry grew, so naturally did the competition. The big gun in pay TV was clearly HBO, yet its share of the market was being whittled away by competitors. HBO had over 60 percent of the pay cable market historically, but by the end of 1981, it had dropped to a shade under 50 percent.

This is how the market had shifted between 1980 and 1981:

• Home Box Office dropped from 60.0 percent to 49.3 percent.
• Showtime grew slightly, from 15.6 percent to 16.4 percent.
• The Movie Channel grew well, from 5.4 percent to 8.8 percent.
• Cinemax moved up sharply, from 1.9 percent to 5.9 percent.

The Money Side of Cable   57

- ON-TV/Oak went up markedly, from 0.9 percent to 4.1 percent.
- Escapade/Playboy doubled, from 0.5 percent to 1.0 percent.
- Spotlight, begun in 1981, had 1.3 percent.

Notice that HBO lost ground significantly, and while Cinemax gained ground significantly, it was not enough to offset HBO's loss. No matter how you measured, the net result for Time Inc. was a lower share of the total market.

At the end of 1980, HBO and Cinemax combined had 61.9 percent of the total pay cable TV market. By the end of 1981, that combined total had slipped to 55.2 percent.

The problem of churning continued to bother the industry a great deal, yet the people in the business didn't seem to be able to get a handle on how to solve it. John Billock, the HBO executive in charge of marketing, said that 35 percent of all new HBO subscribers who are going to cancel the service do so in the first ninty days, and 60 percent disconnect in the first six months. Nationally, HBO's disconnect rate was running about 4 percent a month. That works out to 48 percent a year. This means that HBO had to sign up almost two customers in order to keep one.

With that kind of customer rejection, any other business would have been opening a vein, yet the cable people mostly believed they would just keep on selling more customers, because the number of potential subscribers was limitless as long as new franchises were being granted and more cable systems were being built.

In spite of churn and all the other troubles of the business, everybody still seemed to want to get into cable programming. Consider these two entities from opposite ends of the aesthetic spectrum: the National Geographic Society and *Penthouse* magazine.

In 1981, Dennis B. Kane, National Geographic's director of TV, announced that a consortium of thirty-five nonprofit groups had been formed to provide programming for National Geographic's proposed pay cable service set for 1984. Included in the group were the Smithsonian Institution, the UCLA Film School, the Canadian Film Board, the Wildlife Fund, and the Metropolitan Museum of Art.

At the same time, *Penthouse* publisher Bob Guccione wanted part of the cable action with his Penthouse Channel. He could hardly stand back and let his rival, Hugh Hefner, hog the soft-porn market with his Playboy Channel.

On the other hand, there were some people who wanted to get *out* of the cable programming business. Viacom and Westinghouse's Group W Cable jointly owned Showtime, HBO's main competitor, but people at Westinghouse were starting to get antsy about Showtime's tendency to edge into the R-rated entertainment and decided to bow out. Group W sold its half of Showtime to Viacom.

Westinghouse's plan was to use the money from the Viacom sale to join ABC in competing with Ted Turner's Cable News Network with a new twenty-four-hour news operation to be called the Satellite News Service (SNS); to develop a joint venture with Disney Studios called the Disney Channel; and to buy into the Nashville Network—another joint venture, this time with Opryland Productions (which also owns Opryland Amusement Park in Nashville and broadcasts the Grand Ole Opry).

Westinghouse has always been a maverick in the broadcasting business, often heading in different directions from the rest of the industry. Part of this may be due to the somewhat stiff, straight-arrow leadership it has had, first under Don Gannon and now under Dan Ritchie.

Ritchie wanted *out* of Showtime; he thought that too much tits-and-ass was showing up on the screen, and he was personally offended. What's more, Westinghouse was never entirely sure that it ought to be in the entertainment business at all. (In 1986, it would finally get out of cable entirely, selling its Group W cable system conglomerate to a consortium headed by HBO's sister subsidiary, American Television and Communications, for a whopping $2 billion.)

Meanwhile, RCA's The Entertainment Channel (TEC) was attempting to attract both cable operators and subscribers, trying to connect mostly with newly built cable systems so as to bag those potential subscribers still caught up in the wonder of all those beautiful new channels in the living room. RCA wanted to be on systems with several other pay channels so that paying extra wouldn't seem that unusual and so that TEC's upscale type of programming with its emphasis on foreign films could be seen as sharply different from the other pay channels.

One of the obstacles to TEC's getting on cable systems around the country was the man who was the head of TEC, Arthur R.

Taylor. Formerly the president of CBS, Taylor had spent a lot of time in previous years testifying in Washington and giving speeches around the country attacking cable as an economic pirate and a force for evil. He may have been able to forget what he said about cable, but the cable operators he offended over the years certainly hadn't.

Beyond that, there are some cable systems that are important for a new programming service to be on in order to succeed. A good example is Sterling Manhattan Cable, which serves some parts of Manhattan. Anything on Sterling Manhattan can readily be seen by the big advertising executives who make the important economic decisions concerning a program's survival. If you can't get on Sterling Manhattan, you may be in a lot of trouble. TEC couldn't get on Sterling Manhattan; neither can Showtime or The Movie Channel. Many cynics suspect that the fact that Time Inc. owns Sterling Manhattan may have something to do with all of this.

At the end of September 1982, *New York Times* reporter Sally Bedell did a two-part series called "The Future of Cable TV is Being Fashioned Today." It was about a dream the studios and cable operators have had for several years, called pay-per-view, which means charging the viewer for each movie or program he or she sees. pay-per-view is a critically important issue for the entertainment industry. We'll discuss it in Chapters 29 and 30.

Bedell's opening point was that *Star Wars* had just been shown on some pay-per-view cable channels two years ahead of when it would be on CBS television. Each of the 1.5 million pay-per-view customers paid $8 a head. This was a fortunate experiment because, she said, the industry was finally taking off the rose-colored glasses, abandoning the hype, and taking a look at the reality of cable television economics.

Well, at least *some* people were.

Chapter 13

1982—Excedrin Headache Year

There is no country in the world that devotes as much time, energy, and money to the creation, distribution, and enjoyment of and the involvement in the entertainment business as America. If you want to put a figure on it, try $52 billion—that's with a "B"—and it is more than we spend on anything else but nuclear war. Yet, the entertainment business is volatile, because it is dependent on the all-too-fickle public taste. Cable TV is exciting and profitable, but it can also be depressing and a money drain. It was chiefly a depressing money drain in the early 1980s.

A lot of sour press was hitting the cable business; on Wall Street, the mood was getting increasingly cranky. By mid-1982, investors were abandoning the cable industry. The word in the press was that cable didn't seem to have a direction or a focus, and it couldn't get its act together. This was particularly disappointing to many people because of what cable *could* be if it tried. But the cable industry didn't seem to want to try for excellence; it was interested only in maximum profits. And profit couldn't always be counted on, particularly as the competition grew.

A study commissioned by the cable industry's own National Cable Television Association came up with the conclusion that cable operators were seriously threatened by the over-the-air pay TV competitors, such as Satellite Television (STV), Multiple Distribution Systems (MDS), Low Power Television (LPTV), and Satellite Master Antenna Television (SMATV).

These were all systems for sending television pictures into people's homes without going through the local cable operator's

1982—Excedrin Headache Year 61

system. For example, with SMATV the television signal was pulled off the satellite up in space by a receiving dish located at a hotel or large apartment or condominium development. Then it was distributed to each room, apartment, or condominium by a private cable system contained entirely within the boundaries of the private property. (The only time one needs a publicly franchised cable TV license is when one crosses public property lines. If you keep the whole thing on your own property, no license is needed.)

In most instances, of course, the local cable system operator hates the SMATV installations in his franchise area, because they rob him of a lot of customers. However, some cable operators see advantages to alternative signal distribution methods.

For example, Media General of Fairfax County, Virginia, was granted the franchise to build the cable system in that wealthy suburb of Washington, D.C. The company realized that it was faced with paying out a lot of money at the beginning to get the system built and it would be some time before any revenue would be coming in to offset those expenses.

So Susan Cieslak, an executive at Media General, went out and sold apartment and condominium complexes in their franchise area on SMATV as an immediate interim service until the cable system was built. The plan worked wonderfully; it brought paying customers on line early in the game, because a SMATV installation is quick and easy to build.

Another wrinkle in the cable system is created by people who buy their own satellite receiving dish, called a Television Receive Only (TVRO) dish. The May 26, 1982, edition of *The New York Times* carried one of the early stories about people who have these backyard receiving dishes and pull everything off the satellites whizzing around up there without paying anybody for it. These TVRO owners, who number in the tens of thousands, can get basic cable and HBO and all the other pay TV services without charge. Naturally, this is not good for the cable business, since it means a lot of lost revenue and profit.

Soon after the newspaper story was published, the Babson investment group told its clients that the future of cable did not look all that rosy. The Wall Street gossip circuit moved the message around with the speed of light.

Then, in June 1982, the U.S. Supreme Court stuck it to cable operators in the infamous Loretto case. In this case, an apartment house owner told the local cable company that it would have to pay to run the wires and equipment onto her property. The cable company, in another of those incredible public relations gaffes for which the cable industry has made itself famous, told the apartment house owner to go whistle.

Well, whistle she did to the nine old guys on the U.S. Supreme Court, and guess what? They agreed with her. As of June 1982, building owners could charge cable operators for access to their buildings to service customers who were tenants.

On July 11, *The New York Times* took another swipe at cable with a story on the front page of its business section that blared out, "Tougher Times for Cable TV," and talked about all the competing technologies that were on line or coming on line. The article reinforced the study done by NCTA about STV, SMATV, and the rest.

Stories in *The Wall Street Journal,* the *Los Angeles Times,* and *The San Francisco Examiner* about assorted cable lawsuits, service problems, financial difficulties, and other troubles just made matters worse. All in all, one did not get a rosy portrait of the cable business.

Warner Communications and the troubles plaguing it were indicative of the mood that year. Warner-Amex Cable was in the business of building and operating cable systems in some of the bigger cities around the country such as Cleveland, Indianapolis, Milwaukee, and Dallas. Warner-Amex Satellite Entertainment Company (WASEC) provided programming for cable systems. In fact, it had been responsible for creating some of the most imaginative programming around, such as the Music TV Channel (MTV); Nickelodeon, a channel directed toward pre-teen kids; and The Movie Channel.

However, Warner-Amex Cable and Warner-Amex Satellite Entertainment had experienced some disappointments, and in spite of an aggressive expansion strategy—or, perhaps, because of it—that resulted in the two companies' growing by 70 percent with a combined income of $350 million, the bottom line for 1982 was going to be $60 million in red ink.

Warner had used up almost all of the half a billion bucks it had borrowed from the First National Bank of Boston. The decision was made that the time had come to sell off some of what it had.

One logical candidate for selling was The Movie Channel, because there were several eager buyers for the competition to

HBO and Showtime. The top guy at Warner, Steven Ross, and his people were negotiating to sell to a partnership of Warner Brothers (a brother company), Universal, and Paramount movie studios, all of whom were salivating to get a piece of the cable business.

The talks broke down in September, but it was expected that they would start again in October. Ross certainly hoped so, because Warner-Amex needed the money to pay off debts and keep the expansion program he envisioned going. He felt everyone would benefit from the new Movie Channel partnership because it would ensure a steady supply of movies to the cable service. It would also put the studios into the cable programming business so they could bargain more effectively with their nemesis, HBO.

The studios wanted very much to break HBO's demands for exclusivity and its flat-price offers for movies. They also wanted to get paid on a per-subscriber basis, something HBO continued to balk at.

On Columbus Day 1982, Michael Fuchs, HBO's vice president of programming, told a Hollywood audience that HBO was a firm believer in having exclusive film rights and in paying flat fees for movies. He chided the audience of disbelievers for regarding HBO in particular and pay TV in general as a threat. He told them that pay TV and HBO were a great boon—the savior of the entertainment industry. (Some witnesses said it was like the Judas goat telling the cattle he wanted to lead them up the ramp to the slaughterhouse for a character-building experience that they would remember for as long as they lived.)

While movie studios were eagerly jumping into the cable television programming business, several programming services were cutting back on staff; things were not going as they had hoped. There was a general air of gloom in most advertising-supported cable systems. Earlier, it was projected that the year would see $250 million in ad revenue for cable programming services. As the year drew to a close, it looked as if the number would be somewhere between $150 and $200 million.

Then CBS Cable folded, and everybody's nervous stomachs got even more nervous.

George Maksian of the New York *Daily News* said that CBS Cable was a class act aimed at the nation's discriminating viewers. Still, on December 16, 1982, at 4:30 in the morning, it folded, and no one was quite sure why. Was it entirely due to lack of advertisers' support? In a sense, yes, but it had only been on the air for a

64　Inside HBO

touch over a year. CBS and other networks had been known to carry shows longer than that before they caught on. So what happened here?

First of all, CBS Cable was the personal project of CBS chairman William Paley. When Paley finally retired, CBS Cable lost its mentor and biggest fan. (Paley would return as chairman of the board in 1986 after the ouster of CBS president Thomas Wyman and his attempt to sell CBS to Coca-Cola, but that was far too late to revive CBS Cable.) Second, once again CBS executives forgot the long years they had maligned cable operators as a threat to the American Way, the Canadian provinces, and the generations yet unborn. The CBS execs forgot, but the cable execs did not. Most of the cable people didn't make channel space available for CBS.

So CBS Cable bit the dust, leaving behind as an enduring memory probably one of the funniest party stories in cable history. At the National Cable TV Association convention in Las Vegas just before CBS Cable died, CBS staged the most lavish party imaginable to woo cable system operators.

The theme was "Arabian Nights," and CBS caterers literally went out into the desert near Las Vegas and created a tent-covered oasis with an enormous spread of food and drink, dancing girls, strolling troubadors—the works. Then guests from the convention—there were several thousand—were bused out to the location, and everyone was invited to have a good time.

What the CBS people didn't understand was what happens in the desert as soon as the sun goes down. Two unpleasant—and for CBS disastrous—things happen. The temperature drops very sharply, and the wind starts to blow. Within minutes the thousands of CBS guests were shivering and being sandblasted by the desert wind. People headed toward the buses, but there weren't enough buses to take everyone back to the hotels all at one time. CBS Cable probably didn't make a lot of friends with that party, but it certainly was an experience that few who attended will soon forget.

Elsewhere other cable programmers were cutting back during this dry period—the Disney Channel organization, for example. Part of the reason was the breakup of the joint deal between Disney and Westinghouse's Group W Satellite Communications, another example of Westinghouse's uncertainty about whether it should be in the cable business.

Suffice it to say that the tone of the National Cable Television

Association convention in Las Vegas in May 1982 was characterized by uncertainty. *Barron's Weekly* reporter Gigi Mahon did a long report on the 1982 NCTA gathering, and felt the mood of the convention fitted its location. The cable people, she said, are taking a lot of big gambles, and no one was certain any of them would pay off.

I was there, and I heard one cable exec comment, as he looked around the hall at the bustling exhibits and hucksters, "The year 1982 will be a watershed year for this industry. Fully half these companies won't be around twelve months from now."

Beyond that, the bad customer service continued to be a bugaboo at many cable systems. In a story labeled "Pay Cable TV Is Losing Some of Its Sizzle As Viewer Resistance, Disconnects Rise," the November 19, 1982, *Wall Street Journal* reported that the churn rate was up sharply, as was customer dissatisfaction.

No question—1982 was not a good year for cable.

Chapter 14

TV-Cable Week

Knowing the story of Time Inc.'s most expensive magazine venture in history, *TV-Cable Week*, gives us a rare opportunity to see what makes some executives in Time's media empire tick—and squirm.

Technically, it is the story of a new magazine that failed, but that is not the most important part. It also offers a look at corporate politics and the dynamics of the people who make the decisions at Time Inc., and that has a lot to do with where Home Box Office has been in the last fifteen years, where it is today, and where it will be tomorrow.

The full story about the magazine side of the venture is detailed in Chris Byron's intriguing 1985 book *The Fanciest Dive*, but here we will look at the thinking (or perhaps lack of thinking) that produced the debacle of *TV-Cable Week*.

There is a small cadre of executives who run Time Inc. They share the same visions, worries, and values. The impression one gets from working in the Time-Life Building and from studying its business track record (I have done both), is that the Time Inc. executive team is long on loyalty and a clubby kind of amateurism and self-regard but short on insight and solid business ability.

At one time, for example, Time Inc. got nervous about a possible shortage of newsprint, so it hit upon the idea of buying a forest and paper mill. In the corporate fad of the day, Time Inc. exchanged some Time Inc. stock for the Temple-Inland forest products company of Texas.

It was only after the deal was made that they discovered that the forest and mill they bought were better for making cardboard boxes than for producing the slick newsprint needed for such magazines as *Time, Fortune,* and *Sports Illustrated.*

Other Time Inc. executive goofs include Time-Life Films, *The Washington Star,* two video operations in teletext and satellite television, and giving Austin Furst the home video rights to all of those Time-Life films. There were others, too, such as selling off broadcasting stations to McGraw-Hill just before they became very profitable and trying hard to get out of the cable business at just about the time Home Box Office was being started.

And that brings us to the beginning of April 1983, when the executive suite at Time Inc. launched its most ambitious project in history. It was a new magazine called *TV-Cable Week.*

Multichannel News rang out on April 4th about a computer-generated TV guide that was Time Inc.'s biggest and most expensive magazine launch in history. Time Inc.'s president, Dick Munro, predicted that it would be the biggest and most profitable publication ever.

The way Munro and his crew planned to market this new magazine was a little different from the usual procedure for a magazine. They planned to peddle it to the cable operator and have the cable operator, in turn, sell it to the subscribers. By contrast, its main magazine rival, *TV Guide,* the largest circulating publication in America, got more than half of its sales at newsstands and supermarket checkout counters.

At the time, down in the Washington, D.C., suburban town of Alexandria, Virginia, I was editor/publisher of America's oldest daily newspaper, the *Alexandria Gazette.* I was working to make a local cable TV guide profitable, and I was amazed by the *TV-Cable Week* project. I didn't see how Time Inc. was going to be able to manage it.

There were several major problems I saw. One was that each issue for each cable system had to be individualized. Each cable system in the country has a different numbering arrangement. On system A, for example, the CBS channel might be 2 and on system B, it might be 11, and on system C it might be 39. The other problem was trying to sell the magazine through cable operators, whose primary interest was in running a cable business—*not* selling magazines.

I couldn't see how they could make it work, but I told myself and my staff that maybe some small-town publisher like me couldn't figure it out, but that Dick Munro and all that high-priced talent in New York, working with the deep pockets of Time, might be able to.

Both Dick Munro and I were wrong.

After a couple issues of *TV-Cable Week*, the critics got their claws out. For instance, Bob Brewin of *The Village Voice* said that never before had so much money been spent on a magazine that was so worthless.

He said that other Time Inc. magazines were for people who couldn't read, think, or do either, and he wondered where that put *TV-Cable Week*—which was worse than the others.

Brewin found that, while the magazine promised the reader listings for all channels, this was not what it delivered. In fact, each cable operator could veto the listing of channels he didn't want people to know existed.

Brewin also noted that *On Cable,* a smaller but very successful cable guide, reported that its sales had gone up since *TV-Cable Week* began. According to Brewin, the reason was that many cable operators disliked Time Inc. intensely, regarding the organization as an arrogant, grasping octopus that owned TV pay cable programming services, HBO and Cinemax and part of USA Network. Instead of selling the new Time Inc. magazine, they recommended the competition.

Apparently, Time Inc. and *TV-Cable Week* didn't know anything about the cable operators' antipathy toward them, because they had not made a market survey. They had conceived the magazine inside the Time-Life Building without talking with people outside the Magazine Group. They didn't talk with their potential customers, and they didn't even talk to their colleagues at HBO. For example, Jerry Levin's office was only a few steps down the hall from the office of the group vice president for magazines, Kelso Sutton, but they never talked to each other about this project.

Given the internal rivalry and envy, that probably isn't all that surprising. The Magazine Group desperately wanted a winner that would show it could produce profits just as well as the Video Group did. It probably struck many in the Magazine Group as a delicious irony that they planned to score that triumph with a magazine that was about cable television.

At the same time, Dick Munro was hunkering down to make Time Inc. safe against corporate raiders. Every unit of Time had to do its share by making a good profit, and that included *TV-Cable Week*.

This obsession of top corporate executives to do everything to avoid being taken over is widespread in America these days. Perhaps this takeover phobia will subside a bit now that the Securities and Exchange Commission and the Congress are investigating corporate takeover mania and Wall Street arbitragers such as Ivan Boesky, but it won't go away entirely. It certainly was there in strength in 1983.

That's why Munro wasn't pleased with the July 25, 1983, story in *Communications Daily* that reported that *TV-Cable Week* was in trouble. It quoted the publisher, Daniel Zucchi, as saying the project was a lot tougher than he thought it would be. If you read Byron's book on what was going on at the time, you will know that Zucchi's main problem was tremendous internal conflict and confusion. One gets the clear impression of a project being run like a Keystone Cop movie.

Time's first open move to acknowledge that the project was in deep yogurt was to dump Dan Zucchi as publisher of *TV-Cable Week*. However, in the time-honored tradition of Time Inc.'s taking care of family, it simply shifted Zucchi back to *People* magazine, where he had been before.

The new kid on the block—perhaps the chopping block—was Christopher Meigher, who was transferred over from Time Inc.'s corporate circulation division.

Edwin McDowell of *The New York Times* business section said at the time that it was hard to believe that Time Inc., with seven profitable magazines, was making such a botch of *TV-Cable Week*. The poor judgment and poor management, McDowell said, was giving the financial community the jitters and making Time Inc. stock drop.

On September 16, 1983, Thomas B. Rosenstiel of the *Los Angeles Times* gave the bye-bye story for *TV-Cable Week*. He said that Time Inc. was killing the magazine after five floundering months and a loss of almost $50 million.

Once Time Inc. would have stuck with its projects for years until they turned profitable. For example, it took *Sports Illustrated* ten years to turn the profit corner. However, in the era of corporate takeover phobia, Time Inc. had to keep its profits up so as to keep the price of its stock up—which, in turn, made it more expensive for a raider to buy up the company. Besides, *TV-Cable Week* was different; it was draining money out of Time Inc. faster than a teenage girl with a new charge card. Also, it was distracting top management from other, more important things. Namely, it *was* driving Time's stock price down and making it more vulnerable to a corporate takeover.

One funny quote in the Rosenstiel piece was the reaction of *TV Guide* publisher Erik G. Larson:

> "When they started," said Larson, "I wondered to myself, 'What do they know that I don't know?' I guess the answer is nothing."

The collapse of *TV-Cable Week* gave many people an insight into the mosaic of personalities and cross-currents going on inside Time Inc. Kelso F. Sutton and his magazine colleagues had seen the TV side of Time Inc., particularly HBO, become the most profitable and most important member of the Time Inc. corporate family. Sutton and his print people had wanted to recoup some of their previous glory and saw *TV-Cable Week* as the vehicle to do it. Sutton was also a contender to become president of Time Inc. when Dick Munro retired.

Yet in spite of the magazine's failure, almost nobody important got fired. The Time Inc. country club way of functioning takes care of all the top executives even when they screw up. In fact, some of the key executives involved, such as Sutton, got substantial raises.

Since the debacle of *TV-Cable Week*, Sutton and his magazine people have not been able to come up with another successful magazine project. They have bought a regional magazine, *Southern Living*, and in 1986 they acquired *Science 86*, which they promptly killed in order to help their own ailing *Discover* magazine, which they ended up selling in 1987. In the fall of 1985, Sutton's group began test-marketing *Picture Week*—an anemic imitation of their incredibly successful *People*—in thirteen areas around the country. It flopped, too, and was withdrawn.

Because Munro wanted to be rid of all the losers in his stable, a few days after the *TV-Cable Week* demise, he announced that he was pulling the plug on the company's multimillion-dollar teletext project. (Teletext lets written text be transmitted to home television sets equipped with a special decoder.) As he made that announcement, Munro commented on the death of *TV-Cable Week*, saying that it had been a humbling experience for Time Inc. and that much of the blame was due to "the Time Inc. arrogance."

On October 17, 1983, a postmortem on *TV-Cable Week* by *New York Times* reporter Ian T. McCauley summarized the failure as being due to Time Inc.'s rushing into the marketplace without knowing what it was doing, without having lined up customers in advance, and without understanding how strong the competition was.

Well, aside from that, they had it right.

Chapter 15

The Takeover Jitters

For HBO, as mentioned before, 1983 began with a survey report that married couples ranked HBO #2 in their lives—right behind money and just ahead of simultaneous orgasms. The study, conducted by Dr. Sol Gordon and Dr. Kateryn Everly, was one of the few nonsericus matters that occupied HBO and Time Inc. executives that year.

While the movie studio wars were in progress, HBO programming was turning inward, focusing more upon its own programming efforts than on those of the movie studios. This change reflected several things that were happening in the ever-changing relationship between HBO and its sources of programming.

First, the war between HBO and the studios was making negotiations tougher, so HBO decided it would try to rely less on Hollywood studio movies. In February of 1983, HBO ran twenty-eight Hollywood studio movies, as opposed to thirty-five on The Movie Channel and forty-eight on Showtime. Of course, HBO was in the enviable position of being rich and consequently able to finance its own productions more easily than its competitors.

Second, the Hollywood movie studios weren't turning out enough films to satisfy HBO anyhow. Unfortunately, making movies isn't like manufacturing cars. You can't just crank up the assembly line a little faster to get more of what you want.

And more is not necessarily *better*. Like many creative endeavors, movies are a crapshoot. Everybody is trying to guess what the fickle public will spend money on, and because movies cost an incredible amount of money to make, a great deal is at stake. For example, in late 1986 Frank Price was ousted as president of Universal Studios because of a giant bomb made by George Lucas called *Howard the Duck.* One of the most heavily touted movies of 1986, it turned out to be, as the critics couldn't resist saying, D.D.O.A.—Dead Duck On Arrival. And Michael Cimino's enormous flop, *Heaven's Gate,* drove United Artists to the verge of bankruptcy.

Third, the competition was heating up, and that made Hollywood studio films tougher to get on an exclusive basis—which was still a big deal with HBO. In 1983, Jefferson Graham wrote about it in the Valentine's Day issue of *The Hollywood Reporter,* saying that Hollywood was frustrated by dealing with HBO because it was so powerful and rich. He noted that programming vice president Michael Fuchs was always using HBO's power to pay a lot less for movies than the studios thought he should. This tough bargaining added to the personal dislike that a lot of Hollywood people had for Fuchs.

Still, movies are the name of the game in cable TV. By the beginning of 1983, although HBO had made an exclusive film deal with Columbia, had bought a piece of Orion Pictures, and had joined with CBS and Columbia in forming Tri-Star Studios, it wasn't enough. The cable TV audience yearned for more and more movies.

That's when HBO created Silver Screen Partners with the help of E. F. Hutton, the underwriter, for $100 million. The goal was to produce at least twelve feature films. This was an investment package right up the alley of investment analyst turned HBO president Frank Biondi. Limited partnership shares were sold to the public for $5,000 each. HBO made pre-buy payments of up to 50 percent of each film's cost and guaranteed the remaining 50 percent to the partnership. A no-lose situation for investors, it also attracted a lot more production money to make a lot more films for HBO.

In return for its role, HBO got the exclusive pay TV rights to all of the films for the U.S. and Canada; 25 percent of whatever came in from sale of the film to commercial networks; and the right of first refusal on the film's home video rights. After a film had covered its costs, HBO got the first 5 percent of profits, and the partnership and the producer split the rest evenly.

Silver Screen Partners would continue to be used by HBO because it was imaginative and very attractive to investors, while at the same time it beefed up the supply of programs available to HBO. And notice that, even though this was 1983, before everybody got overly excited about the videocassette recorder and the home video market, HBO was locking up the first-refusal rights for home video. This tendency to look beyond today's market is one of the things that has kept HBO in its dominant position in the entertainment business.

Although HBO *was* doing well, the men in the executive suite in 1983 were growing increasingly nervous about the possibility of a takeover. People on Wall Street saw Time Inc. as a juicy item for a corporate raider. Time Inc. executives knew that if the company were taken over by an outsider most of them would lose their jobs, and that made them really nervous.

For example, cash-heavy Coca-Cola had taken over Columbia Pictures and was aggressively looking for other expansion possibilities. It could be a smart move for Coca-Cola to take over Time Inc., revamp management, and dump the print properties. Knowing that this sort of thing was a definite possibility, Munro moved to make Time Inc. less vulnerable.

Munro's idea was to spin off the Forest Products Division, because, while it was profitable (in 1982 it made $55 million profit), it was not as profitable as the Video Group, which made $166 million profit in 1982. Besides, both operations needed heavy capital investments for 1983–84, and the Video Group seemed to earn a better return. The plan was to spin off Forest Products to current Time Inc. stockholders and buy out the 3 million Time Inc. shares controlled by Art Temple, the main man in the Texas forest products group.

Munro made other moves to protect against a takeover, such as getting the stockholders to allow management to reincorporate Time Inc. in Delaware, where corporate laws are more favorable in protecting management from outside threats than in most other places inside the United States.

He got the shareholders to agree to management's buying back some of Time Inc.'s stock on the open market. The purpose of this was to cut down the number of stockholders and the number of outstanding shares. That, in turn, reduced the number of "strangers" who simply invested in the company but had no particular company loyalty and were likely to sell out to the highest bidder. When one cuts the number of outstanding shares, but keeps the same assets in the company, it normally makes each share more valuable with a higher market price. This, in turn, makes it more expensive for an outside raider to buy enough to have a good run at gaining control.

As we saw earlier, Munro was deeply concerned about which divisions of the company were winners and which were losers. He wanted to push up the company's stock price to make it more expensive for any corporate raider. There was constant pressure either to make the losers into winners or to dump them.

Munro had to watch out for losers like Teletext, Satellite TV, and Time-Life Books, which had made a $78 million profit in 1978 but was struggling to break even in 1982. With STV, Munro had already sold the system Time owned in Dallas, but when he couldn't find a buyer for the one in Cleveland, he just pulled the plug on it and shut it down.

In an attempt to improve management, Munro had Time Inc. managers spend some pleasant days at the end of 1983, in Palm Springs. As idyllic and low-pressure as the desert is at the foot of Mount San Jacinto, the message was clear: *Manage better. Plan better. Get your returns up.*

Laura Landro, the *Wall Street Journal* reporter who wrote about the Palm Springs meeting, summarized the essence of what went on in the December 6, 1983, issue. Among other things, she said that the management of Time Inc. had never needed to learn how to manage better more than it did then. She said that *TV-Cable Week* had gone belly-up, and subscription TV and teletext projects had to be dropped as losers. Still, several other things had to be done to bring the company up and fend off potential takeover raiders; two necessary steps were creating a more centralized management and dispensing with the country-club culture at Time Inc. Finally, she noted that not all was well in HBO. She mentioned a 36 percent annual disconnect rate and the strong threat of home video. These were telling points.

One of the things Time management did at this point was to hire an outside consulting firm, McKinsey and Company, to come in and tell Time Inc. what it was and where it should be going. Another telling point.

Chapter 16

HBO Goes Hollywood and Hollywood Goes Cable

Almost since HBO began, it has been in a running battle with Hollywood's movie moguls, and always over the same issue: money. But these battles have generated a level of hatred, anger, and venom that goes beyond the limits of ordinary business negotiation. The fighting is vicious and cutthroat, and vengeance and vendettas are the order of the day.

This may be because of the personalities involved or because the business is unpredictable and volatile and mistakes can cost millions. Or perhaps it's because each side keeps trying to cover itself by edging into the other one's business. From 1981 on, HBO has been pushing very hard to get into "original programming." And the studios have flirted with getting into the cable business, even before HBO was born.

But in spite of bickering and contradictions, Hollywood and HBO have always needed each other. In 1981, Jerry Levin, who was then vice president of the Time Video Group, told the English magazine *Intermedia* that sometime in 1982 Hollywood's income from pay cable would equal half its income from theaters. When it went to twenty-four-hour programming, HBO needed about 550 movies to run each year, a substantial jump from the 280 movies it ran in 1978. Levin also said that HBO could not survive without a healthy movie industry.

HBO Goes Hollywood and Hollywood Goes Cable 77

HBO needed Hollywood to produce more movies. To stimulate more production, HBO put together a partnership in 1982 with Columbia Studios and CBS to form a new studio, Tri-Star Productions. Tri-Star was partly the brainchild of Victor A. Kaufman, an attorney who was an executive vice president at Columbia at the time and who understood that the three most important things for the success of a movie were distribution, distribution, and distribution. If it isn't out where people can see it, it will not be a success.

Kaufman calculated that the three major avenues of distribution for a movie were movie theaters, commercial television, and cable television. By forming a company that had access to each of these three, Kaufman could be sure that any movie it produced would get wide distribution and be a financial success. As we will see in a later chapter, the three-way partnership would change between 1982 and 1988, but it would be an astonishing success along the way.

It was ironic that CBS, which had spent years fighting cable TV as a menace, was now joining cable in a corporate marriage, and that HBO, which had fought with both commercial television and movie studios, was to become a partner with both. It's a strange business.

Actually, Time Inc. had been briefly involved in the movie-studio business before. In 1979, Andrew Heiskell, Time Inc.'s chief executive officer, decided to buy out David Susskind's movie company, Talent Associates. Unfortunately, Time's early skittishness in buying into a movie studio was justified by the bewildering and frustrating way that Talent Associates was run. In 1981, Time Inc. gave up and sold out.

The whole thing left a sour taste about the movie business in Heiskell's mouth. He discovered too late that he disliked just about everything about Hollywood, from the pretentious display of wealth to the unorthodox methods of conducting business.

On the other side, in spite of all the badmouthing of the cable business by Wall Street, the critics, the networks, and assorted other naysayers, the movie crowd was still fighting to get into cable business. They wanted in because they saw all the money that HBO was making.

For example, Paramount, Universal, and Warner Brothers wanted to buy 75 percent of The Movie Channel (TMC). The problem, as usual, was that the studios were more afraid of one another than they were of anybody else. Still, there was the strong motivation to get *into* cable and to get *at* HBO.

Why was everybody out to get HBO? The obvious answer was that HBO was arrogant and abrasive—something for which it would pay over and over again in the years to come. What's worse, it was smashingly successful and made lots of money.

One of the most aggravating things that HBO did to the studio chiefs was to make "pre-buys"—putting money into an independent producer's hands *before* the shooting began. That meant that none of the producers had to grovel before studio heads or bankers for the money needed to make a movie.

In exchange, HBO got the exclusive pay TV rights to that movie, which might then be offered for theater distribution by the producer to one of the movie studios. However, while that studio could get part of the take from the theaters, it was frozen out on any pay TV rights because HBO already had them.

Predictably, the HBO pre-buy strategy outraged the movie studios. Also predictably, the HBO pre-buys didn't always turn out to be successes. Consider *The First Deadly Sin* and *The Legend of the Lone Ranger*. They were not exactly huge successes, but they cost HBO about $3.5 million.

On the other hand, HBO made big pre-buys in *Sophie's Choice* with Meryl Streep, *High Road to China* with Tom Selleck, *Daniel* with Timothy Hutton, *First Blood* with Sylvester Stallone, and *The King of Comedy* with Robert DeNiro.

To make the pre-buy for *The Pope of Greenwich Village*, HBO anted up a reported $7.5 million, and for that it got *all* rights beyond the theater run—pay TV, home videocassette, and even the commercial TV rights. Again, HBO was looking beyond pay cable.

The way this worked out financially was not too complicated. While many studios were trying to sell their blockbuster movies to cable TV for a price of $1 per subscriber, HBO actually paid less because the number of its subscribers was growing. Take the case of *The Pope of Greenwich Village*. HBO paid $7.5 million up front. It took a year or two for the movie to be shot and edited. Then it took another year for it to be shown on the theater circuit. So, two years after putting up the $7.5 million, HBO had the movie to show to what by then had grown to 15 million subscribers, at a cost of about fifty cents per subscriber.

Beyond that, HBO would pick up more money by selling the

movie to the home videocassette and the commercial television network markets. This is what prompted one studio executive to say that never before in the history of the entertainment business has so much power been concentrated in a single company—HBO.

This is also why the studios kept trying to put together a deal that would give them a piece of the cable business. As the October 21, 1982, *Wall Street Journal* noted, the movie studios were very anxious to get into pay cable TV to blunt the power of HBO and other pay cable TV services, such as Showtime and The Movie Channel.

By mid-November 1982, three studios got together to give it another try. Paramount, Universal, and Warner Brothers tried to buy a piece of The Movie Channel from Warner-Amex. Collectively, they were putting up $75 million for 87.5 percent of The Movie Channel, which was better than the earlier negotiations, in which Warner-Amex wanted $75 million for 75 percent. Of course, the Justice Department would look at the deal carefully.

Meanwhile, ABC was talking to Viacom about buying a piece of Showtime. Also talking to Viacom were Columbia Studios and 20th Century–Fox. Even though Showtime had only 3.5 million subscribers, compared to HBO's 11 million, it was an attractive service and offered a way for the studios and ABC to get into cable.

Jane Hall, editor of the publication *View,* said that the studios saw Showtime as the last boat leaving for Pay Cableland, and they wanted to be on it. The boat would take them, they believed, to that paradise called pay-per-view, pay cable, and home videocassettes.

All this wheeling and dealing produced an air of intrigue in the business as confidential calls were being made and secret meetings held all around Manhattan and Los Angeles.

For example, when the CBS-Columbia-HBO deal on Tri-Star was made, it dramatically killed another major deal that was quietly being negotiated: for Columbia to buy into Showtime along with ABC. This other deal was supposed to be wrapped up at 8:30 A.M. on November 21, 1983, in the Sixth Avenue offices of ABC in New York. It was supposedly a done deal; all that they needed were a few signatures on the dotted line, and ABC, 20th Century–Fox, and Columbia Pictures would own 75 percent of Showtime.

By the 8:30 starting time, most of the principals had gathered at ABC headquarters. There was Michael P. Mallardi, American Broadcasting Company's chief financial officer; Herbert Granath, president of ABC Video Enterprises; Burton Monasch, Fox's executive vice president; and Steve Roberts, president of the telecommunications division. The only person missing was the man from Columbia, Victor A. Kaufman. When he hadn't shown up by 9:00 A.M., somebody telephoned him. When he refused to take the call, Roberts and Monasch got angry, marched over to Kaufman's office, and bullied their way in.

That's when they found out that Columbia, which had just been bought by Coca-Cola, had snookered the rest of them by making a secret deal over the weekend with HBO and CBS to create Tri-Star. Insiders say that the factor that solidified the HBO-CBS-Columbia deal was a personal visit by Time Inc.'s Dick Munro to the home of Thomas Wyman, the number-one man at CBS, over Thanksgiving weekend.

Hollywood–New York entertainment relationships can be as intertwined and as complicated as Hollywood marriages. It reminds me of the two kids at a Beverly Hills elementary school. One was bragging, "My dad can beat your dad." The second one replied, "What are you talking about? My dad *is* your dad."

PART III

CABLE TV ISSUES

Chapter 17

T & A on Cable

The hair on my arms was standing on end. There it was. Nudity. Naked bodies. Sex. On television.

It was the summer of 1983, and Diane Galuski of Buffalo, New York, was telling *The New York Times* about her unexpected encounter with the Playboy Channel on her cable TV.

Galuski wasn't your average viewer, though. She was the local chapter president of Morality in Media, a Catholic media watch organization founded more than twenty years earlier by a Jesuit priest. She said that this was what Morality in Media had predicted would happen—sex would move from the back alleys to the porno shops, and finally into the living rooms of America.

One year after Galuski saw those nude bodies on television, the Buffalo City Council, acting on a petition with eleven thousand signatures, voted on whether or not to ban the Playboy Channel from local cable TV. The vote was taken just before Halloween of 1984 amidst much regional publicity. In the end the city council turned down the ban by a vote of ten to three.

Later, city council president George K. Arthur reported that he had watched the Playboy Channel at a friend's home. While he found the shows boring and disgusting, he was against censorship. He did not like the idea of one group imposing its moral standards on another group.

While Diane Galuski was being uptight about the Playboy Channel, some four thousand of the sixty-three thousand cable subscribers in Buffalo were not, and they were sending in $10 a month to see tits and ass on cable.

Some of them noted that if Galuski and others didn't like the

84 *Inside HBO*

Playboy Channel, they didn't have to subscribe to it. After all, it was a special pay channel, and it must be specifically ordered.

The Playboy Channel ("We take the staples out of the centerfold" was its slogan) has been the target of sin-fighters ever since it began in 1982. It has been a subject of controversy in every city in which it has been introduced into the local cable system. The Channel was banned in Cincinnati and in Memphis after hundreds of cable subscribers complained about movies such as *Love and the Sensuous French Woman*.

When it began, the Playboy Channel was a joint venture of Playboy Enterprises and Cablevision, in Woodbury, New York; Daniels and Associates in Denver, Colorado; and Cox Cable Communications in Atlanta, Georgia. Cablevision is the Chuck Dolan enterprise. Daniels and Associates is a Denver-based brokerage firm that deals in the buying and selling of cable systems. Cox Cable is a conservative media conglomerate that owns TV and radio stations, newspapers, and cable systems.

The National Cable Television Association (NCTA) says that only a handful of the more than five thousand cable systems show X-rated movies (curiously, one of them is in Allentown, Pennsylvania, in the heart of the Amish countryside). Not even the Playboy Channel shows X-rated programs. It edits its movies down to a hard R-rating.

Playboy may claim to tone down its material somewhat, but *TV Guide* noted that this objective falters in such programs as "Playmate Playoffs." In this mock-athletic event, Playmates in bikinis romp on Hugh Hefner's lawn in Holmby Hills in a frantic effort to win a few prizes while almost staying in their tiny bikini tops.

Home Box Office says that some 40 percent of its movies are R-rated. It explains that by saying that research indicates that their subscribers want even *more* R-rated movies and uses Bo Derek's *Tarzan, the Ape Man,* which was one of the most popular films, as an example.

Cable operators rationalize all of this by saying that their programming does not reach anybody over the free airwaves. One must specifically subscribe to a cable service, and even more specifically, subscribe to and pay extra for such adult services as the Playboy Channel. Therefore, they claim, they are not a public nuisance, nor are they publicly pandering or publishing obscenities.

Some people in the cable business have claimed that the First Amendment gives them the right to put whatever they want on their cable channels. That's certainly the HBO position, proclaimed by its executives, such as its president at the time, Frank Biondi.

(A side note about Mr. Biondi's devotion to the First Amendment. It was on his orders that Jefferson Graham, then a reporter for *The Hollywood Reporter,* was banned from covering one of Biondi's speeches. The subject of the speech: cable TV and the First Amendment.)

Of major concern when I was at HBO was this so-called "First Amendment Issue," but not because it involved a constitutional matter. Instead, HBO was afraid local and state obscenity laws would force it to transmit many different versions of the same program. This would be enormously expensive, and this cost—not the Constitution—was the main worry.

Battles were being waged all over the country. After the mayor of Miami visited New York City and saw cable TV with the movie *Midnight Blue* and another program with six nude men and women involved in sensuous talk and touch action, he got an antiobscenity law passed in his town. It was promptly challenged by cable subscriber Ruben Cruz, and HBO also sued to nullify the ordinance.

The federal district judge, William Hoeveler, upheld HBO's and Cruz's challenges in 1983. He said that the law was "well-meant" but that it violated the Constitution:

> Cable provides greater overall viewing control to the subscriber . . . is totally up to the user to decide to bring Miami Cablevision into his home.

(Not satisfied with the ruling, HBO also sued the City of Miami to recover its legal costs. That wasn't settled until September 1986, when HBO was awarded $50,000 in attorney's fees.)

A lot of anti-obscenity citizen activity has gone on in Utah during the last several years—particularly against cable TV. The Utah Association of Women submitted a petition to the state legislature with sixty thousand signatures calling for a ballot initiative forbidding obscenity on cable TV.

Even though a 1981 state law banning obscenity on cable was thrown out by Utah judges, the legislature passed another law, the

86 Inside HBO

"Cable Television Programming Decency Act," in 1983. The governor, Scott M. Matheson, vetoed the law.

Beyond that, the citizens of Utah voted on the ballot initiative proposed by the Utah Association of Women. It would ban "indecent material" from cable TV. Initiative A, "The Cable TV Indecency Act," came up for a vote on November 6, 1984, and the people of Utah voted it down by a two-to-one margin. The actual vote was 85,305 (33 percent) for banning indecent material on cable TV and 173,889 (67 percent) against any ban.

One of the problems of this issue is an anomaly in the franchising process. Almost all communities demand a "local access channel" when they award the cable franchise in their town. The innocent idea of this is that it would allow anybody in the community to buy air time. Presumably, this would allow local people to show their kids tap dancing to "Lady of Spain" accompanied by the electric sitar.

However, community access channels also give rise to programs such as the one starring Ugly George and his nudie cuties. Ugly George pays Time Inc.'s New York City cable system, Manhattan Cable, for air time, and the cable system is obligated to sell to him. Ugly George then goes out on the streets of New York City with his portable TV camera and talks women into coming back to his place to take off their clothes on live TV. An astonishing number of women do it.

So, what to do? Reverend Don knew.

Donald E. Wildmon is a Methodist preacher who took charge of the National Federation for Decency in Television in about 1979. Out of this group came a new organizing action that would have them quaking along Sixth Avenue in big New York City: The Coalition for Better Television.

Headed by the Reverend Don, The Coalition for Better Television was armed with a simple but terrifying plan. They were going to watch a lot of television for a few months and then organize voluntary boycotts of sponsors who underwrote shows with too much sex, profanity, or violence. Not surprisingly, the idea caught the imagination of anti-ERA crusader Phyllis Schafly and Dr. Ron Godwin, chief operating officer of the Moral Majority, as well as two hundred other pro-life, pro-family groups who heard about it—usually by way of TV news broadcasts.

Naturally, the television executives on the island of Manhattan responded like wounded warthogs. An NBC spokesman said that monitoring television is okay, but "boycotts and other pressure tactics raise entirely different questions." Which, if you analyze the statement for about ten seconds, is a bunch of words strung together saying nothing short of the obvious. ABC didn't piddle around the namby-pamby way NBC did; it said the approach was "a totally unacceptable means of trying to influence programming." (At least it was totally unacceptable to ABC and others in TV because we were talking about money now—big globs of money.) CBS called it censorship, and Roger Shelley, a Revlon vice president speaking as an advertiser on some of the shows that the Coalition had targeted as jiggling too much, responded, "People setting standards for everyone else to live by—that's dangerous."

He ignored the question of whether or not his company was setting the standards of what would appear on television while objecting when someone else wanted to do the same thing. Of course, he was *paying* for the opportunity to say his piece on TV. Perhaps he equated having money with being morally correct.

In fact, that is the view of a surprising number of Americans; namely, if it's just business, there is no moral issue. A prime example occurred during President Reagan's November 19, 1986, press conference when he was defending the sending of arms to Iran. The point he kept making was that it wasn't as if he ordered the arms *given* to the Iranians. The Iranians had *paid* for the arms, which removed it from the arena of morality. But, back to important matters, tits and ass on cable television.

Reverend Don continued to articulate the Coalition's position: Citizens have urged and pleaded with the networks to cut out the blatant sex and violence, but the networks have arrogantly ignored the people.

He said that the networks used the money TV earned to enhance their clout in the community and now these same networks were objecting to other people fighting back in the same way.

Reverend Don was not the senior veteran crusader in this battle. That title would have to go to Father Morton A. Hills, a Catholic priest, former head of the President's Commission on Obscenity, and the current president of Morality in Media. He has been fighting on the obscenity in media battleground for a quarter of a century.

His tactics are focused on cableporn and include such techniques as a letter sent out by former Dallas Cowboys quarterback

Roger Staubach, who asked all recipients to send a postcard of protest to the mayor of their city.

In part the letter warned:

> What I am about to tell you is so shocking, so threatening that you may not believe it!
>
> Homosexual acts, women being brutally molested and raped and other explicit sex acts can now explode into your living room on some public access cable TV channels.
>
> And, if cable TV hasn't reached your community yet, you can be sure it will soon. Don't think this isn't your problem.
>
> The smut peddlers have moved in and cable TV could become a dangerous weapon that might be used against our children and grandchildren.
>
> The great medium of cable TV is being used to "pimp" children. It is showing explicit movies of women tied and beaten, raped with guns and other deviant sexual acts, all of which claim to be entertainment.

How big a deal is this with the general public? We know that a lot of folks in Utah and elsewhere are concerned and we respect that concern. However, even more people are concerned about censorship and the imposition of one person's morality on others who may not want to accept it.

And there are a lot of people who actually *like* cableporn. For example, there is the cable system in San Jose, California, the heart of Silicon Valley, where the Gill Cable system has ninety-six thousand subscribers of which about ten thousand pay extra to get an "adult" movie channel. Of the ten thousand, only three hundred have asked for and gotten the free "parental lock" device.

In Lowell, Massachusetts, ten thousand subscribers took HBO, which does show soft R-rated movies, while only twenty-five hundred opted for the family-fare alternative, Home Theatre Network.

One of my researchers gave me this report:

> Children, undoubtedly, will see naked bodies on adult TV, too, if the experience in Borger, Texas, means anything. **Escapade,** later renamed **Playboy Channel,** was started there in April of 1982.

A year later it was learned that kids at Crockett Elementary School, in this Panhandle town of 14,500, were discussing what they had seen on **Playboy.** "They were talking about it explicitly," said Karen Sterrett, president of the Crockett PTA.

The PTA sent a note home with the children, asking parents to write to the local commission that advises the city council on the cable franchise.

The commission got more than 800 cards and letters—180 opposing the channel, 638 in support.

Even in Texas.
Today, the battle still goes on. Certainly, there is more close-to-explicit sex on TV than ever before. Those who doubt that clearly have not watched "Dallas," "Knott's Landing," "Falcon Crest," or any of the soap operas or movies playing day or night on commercial TV. This is a direct ploy to lure viewers at a time when commercial television's share of the viewing audience is steadily dropping because of cable and home video.

Likewise, the pay cable services such as HBO and Showtime continue to show uncut R-rated movies as their weapon to attract subscribers who can easily get X-rated and XXX-rated movies from the local home video store.

The Meese Commission report on pornography expressed the concern of conservatives over the issue, but was largely a joke typified by the fact that the most pornographic publication of the federal government today is the Commission's own report with all of its exhibits of pornography. There is also the ironic photograph of Attorney General Ed Meese III, holding a public press conference to issue the report standing in front of a towering statue of Justice—a bare-breasted woman.

Chapter 18

What to Do about Home Video?

> The VCR will do to the cable business what the cable business has done to broadcasting.

That was the view of Harold Vogel, an industry analyst with Merrill, Lynch, as quoted in *The Washington Post* in early 1984.

Clearly, the home video explosion caught Home Box Office and a lot of cable people by surprise. Having been blindsided once more, what should cable and Home Box Office people do to counter the impact of home video and the video cassette recorder on the pay TV market?

Here we'll look at it the way it was when the impact of home video first became evident. A little later, we will review the situation as it stands in 1987.

That 1984 *Washington Post* story reviewed what home video and the VCR were doing to the pay cable industry, including HBO. It said it was seriously undermining the whole pay TV industry with the result that pay cable growth in the first quarter of 1984 had dropped to less than half of what it had been the year before. Notice that it was not talking about a *loss* in cable subscribers but rather a *decline in growth*.

Echoing the point made in the *Post* story, Michael Fuchs agreed that home video presented a serious challenge to the pay cable services. A *New York Times* story of February 12, 1984, quoted him as saying that home video was a problem because more than 75 percent of the pay cable subscribers would have VCRs by 1990.

What to Do about Home Video? 91

At the heart of the problem are movie release schedules. New movies go first into the theaters; then to home video; next to pay cable; next to commercial TV; and, finally, to the revival theaters. In other words, home video gets movies months before HBO does. Why, then, should anyone pay money to subscribe to Home Box Office to see a movie when the movie can be seen on videocassette from the local video store four to six months earlier? It's a good question, to which there are no perfect answers.

Before we explore that any further, let's answer a more fundamental question. Where did the tidal wave, or *tsunami,* of home video that is now flooding American homes come from? From the same place the word *tsunami* came, Japan. Incredibly, Japanese industry has in many cases taken an American invention and produced the best-quality, lowest-cost version of it. That's what happened with the VCR.

The videocassette recorder was developed in the United States in 1961, but hardly anybody did anything with it here. However, thousands of miles away the Japanese took it home and studied it. Fourteen years later, Sony came out with their version of the VCR—the Betamax—at around $1,200 apiece.

At that price, it took Sony three years to sell it to 1 percent of the households in America. That worked out to be about eighty thousand VCRs sold in three years. Not great, but not bad, either.

But something was happening out there, and a lot of people didn't realize it. It took three years to sell 1 percent of American householders on buying a VCR, but in the three years that followed, there were VCRs in 3 percent of America's homes. VCRs were catching on, and the majority of companies making them were and still are Japanese or Korean.

By 1982, the market exploded. Between May of 1982 and May of 1984, the VCR homes in this country jumped by 300 percent to a total of 10 million—13 percent of all TV households. By the middle of 1984, *The Home Video and Cable Report,* citing figures from the Electronic Industries Association, reported that VCR sales were running 98 percent ahead of 1983. By the end of 1985, there were more American homes with VCRs than there were American homes connected to cable TV.

On May 27, 1984, *The Washington Post* predicted that home video would change home entertainment as much in the 1980s as television had in the 1950s. It went on to say that home video dramatically changed not just the broadcasting and cable TV industries but the advertising industry as well, because it meant that people would be watching programs without commercials.

An A.C. Nielsen survey for the year ending May 1986 showed that the average VCR owner was taping almost three hours of programs every day, usually during evening primetime or daytime soap operas. Since the average American watches TV more than six hours a day, it would seem that about half the TV-watching of VCR owners could be of programs they had previously taped.

As it developed, home video was also going through its own growth changes. In 1980, porno tapes were what was moving. Then TV series, movies, children's programming, and instructional tapes began to be popular with VCR owners. More recently, how-to-do-it tapes have become popular. For example, one of the hottest video tapes in recent times has been *Jane Fonda's Workout*. Also in demand now are extended-play music videos, which follow the tremendous success of Music TV (MTV). For example, the MTV production of *The Making of Thriller*, the Michael Jackson smash album, sold eight hundred thousand videocassettes.

Roll that number around in your head for a moment—800,000 videocassettes at about $30 profit apiece. That works out to be $24 million from the sale of *one videocassette*. The videocassette market as a whole did $1 billion in 1983 and is predicted to do $5 billion in 1988.

Several years ago, in 1976, the movie studios, led by Walt Disney and Universal, decided that the people recording their movies off the TV set were committing a crime. They sued the Sony company over the making and selling of those infernal Betamax VCRs. The movie moguls wanted a tax on recorders and blank tapes that would be paid to the copyright owners (usually the studios) of the films that a VCR owner might tape.

The case went on for almost eight years, starting in when there were about fifty thousand VCRs in America. Finally, the United States Supreme Court ruled against the movie industry and said that people could tape-record programs in their own homes for personal use without paying anybody anything.

Some legal experts who have followed this case observe that technology is on fast-forward and Hollywood and the courts are on slow-motion. Another historian remembered that once upon a time, the movie studios in Hollywood also wanted Congress to make television illegal.

Channels, one of the better cable-industry journals, com-

What to Do about Home Video?   93

mented in its March/April 1984 issue that even though Hollywood was complaining about home video, it would probably turn out to be the best thing for the studios since television. And, in fact, the studios finally got into the home video business themselves, but as usual, they did it in a strange way. They were forced into it by the film pirates—the ones who made illegal copies of popular movies such as *Raiders of the Lost Ark* or *Silkwood* and sold them for $1,000 each—the studios figured that if they could sell the videocassette rights to someone who would then offer the videocassettes to the public at a relatively cheap price, they could effectively kill the market for the pirated stuff.

So in 1978, 20th Century–Fox sold the videocassette rights to fifty movies for $6,000 each to Andre Blay, a man who had more faith in the home video market than 20th Century–Fox did. Blay peddled the videocassettes to the public for $79.95 each, but his market was quite small. Remember, at that point the VCR was in only 1 percent of America's homes.

Soon thereafter a few other videocassette dreamers moved into the business and gradually got a firm grip on marketing and distribution. In time these companies, such as Commtron and Sound Video Unlimited, locked up a lot of the distribution, and now the studios pretty much have to sell their videocassette rights to one of them. Commtron is doing about $125 million in business a year, and Sound Video Unlimited pulls in around $90 million.

Of course, studios could have had this business all to themselves, but that was not the way of the movie mogul mind. Alan Hirschfield, not considered a mogul by some of his fellow Hollywood honchos, but nevertheless the former head of Columbia Pictures and 20th Century–Fox, summed it up:

> The film companies through the years have failed to have much foresight about the future. We didn't do it with TV. We failed with cable, and we allowed others to seize the distribution of home video.

What could Home Box Office and the rest of the cable industry do about the sharp new competition from home video? Well, Home Box Office could buy and did buy the home video rights to everything they could as part of a package deal from studios and independent producers. In other words, HBO wisely avoided fighting home video and decided to join it.

As far back as 1984, HBO began to search for a home video company to buy into just as it had done with program production companies. Of course, this move had to be made diplomatically, so as not to scare HBO's main distributors, the cable system operators.

One logical candidate for HBO to buy into was Vestron Video, the firm of an ex–Time Inc. executive, Austin Furst. Another was the British company with which HBO was already in partnership on other projects, Thorn EMI.

As usual, Austin Furst was playing it close to his vest. Furst was not interested in a buyout. He wanted his company to remain independent of any conglomerate. He had gotten off to a smashing start because his buddies at Time Inc. gave him the home video rights to two hundred films in the Time-Life Films library (including *Fort Apache, The Bronx* and *Cannonball Run*) for peanuts when he left Time Inc. Coupled with the smashing success of Michael Jackson's video *The Making of Thriller* which he marketed at enormous profit. Since he left Time Inc., Furst has built Vestron into a giant of the home video business. We'll hear more about him later.

On April Fool's Day of 1984, *The New York Times* revealed that all the shillying and shallying were over and HBO was definitely in the home video business. It reported that from then on, consumers would be seeing HBO films not just on cable but in theaters and on videocassette as well. (One of the ironies of the HBO film production and HBO home video setups is that, in order to follow the industry pattern that makes sense anyhow, HBO-financed movies will be available first on home video.)

Three months later, Paul Kagan's *Pay TV Newsletter* of June 29, 1984, reported that HBO had made a deal with Orion Pictures, a studio that HBO helped to get started with a $10 million investment and a prime source of films for HBO since 1982, for domestic home video rights on thirty-three upcoming films. HBO and Thorn EMI finally did announce a deal that brought them together. As for Austin Furst and his Vestron Video, he continued to do fine. In September of 1985, Furst took his company public. Although he was not able to sell as much of the initial stock offering as he had anticipated, he made a multimillion-dollar personal fortune out of the deal.

Chapter 19

A Billion-Dollar Super Bowl and Jail for Pirates

The two things that get a cable operator's heart going pitter-pat faster than anything else are pay-per-view and piracy. Pay-per-view (PPV) is the great golden dream of the future, with untold riches flowing in. Piracy is the dark disaster of the present, with untold riches flowing out.

With pay-per-view (PPV) cable subscribers are charged for watching each performance of a particular program. Many people have experienced a version of PPV in hotels where they may order a recent movie shown on the TV set in their room for a fee—usually $5 or so. With pay-per-view you contact the office of the cable system you subscribe to, and they will transmit the movie or special event you chose on a special cable that only you and others who order it can view. An additional charge is then put on your monthly bill.

PPV is the big talk in cable these days, especially in sports circles and at the studios. One cable executive, Dan Aaron of Comcast, told a Financial Analysts Federation seminar in St. Louis that pay-per-view was going to be cable's next big money-maker. He predicted that 50 percent of the country would pay as much as $15 apiece to see special events—the Super Bowl, for instance—that could mean over $1 billion.

In its October 22, 1984, issue *Cablevision* magazine's special report on PPV noted that this time the studios had no intention of being left out of "the world's greatest theatre." It said that the Hollywood moguls had learned from their mistakes with television; this time they'd be at the front of the line when it came to pay-per-view. It said that the movie studios were looking toward the day that they could release a blockbuster movie to cable TV for a single run on the same day they release it to the theaters.

A test of that theory came on February 18, 1983, when Universal Studios' *The Pirates of Penzance* was released in sixty-two theaters around the country and offered to cable TV subscribers with pay-per-view equipment. This Linda Ronstadt movie was seen by about a hundred thousand pay-per-view cable subscribers, each of whom paid $10 for the privilege.

Some thought it was a disappointing first try, but Mike Marcovsky, a Los Angeles pay-TV consultant, said that it was just the wrong film at the wrong time. It was not a general-interest film, and there were relatively few people with pay-per-view equipment. Even so, it pulled in $1 million in one night from cable TV.

While there is a lot of confusion about PPV equipment and while cable operators are reluctant to invest the piles of money necessary, when put it in place, PPV may be the cable operators' best defense against the home video business. *Variety,* reporting on the Western Cable Show in Anaheim on December 21, 1983, made the point that cable operators needed new services to bring in more money. PPV was seen as the logical answer.

Bob Wussler, with Ted Turner's company in Atlanta, predicts the day is coming when cable subscribers will be paying as much each month for cable TV entertainment as they now do for their cars.

To get some idea of the bucks involved in the PPV dream, industry analyst Paul Kagan foresees the day within three years when there will be 20 million pay TV homes with the necessary equipment to receive a PPV movie. If the subscriber is charged $5 to see a movie on the same day it is released to neighborhood theaters and only 10 percent of the subscribers do so, it could bring in $10 million in one day! Assume a fifty-fifty split between the cable operators and the studios and that means the movie studios would make $5 million. That's about half of what most pictures make from their entire run in the theater circuit.

Bill Daniels, whose Daniels and Associates owns and operates a number of cable systems around the country, doesn't look toward movies on PPV. "The big bucks in pay-per-view are going to be in sporting events," he says. Daniels isn't the only one who believes that the future of pay-per-view is in sports. Warner Communications bought 49 percent of the Pittsburgh Pirates, Group W was dickering for a piece of the San Francisco Giants, and American Broadcasting Company bought 100 percent of the Entertainment and Sports Programming Network (ESPN) to position themselves for the eventual PPV sports market.

Fight promoters Bob Arum and Don King were already setting up PPV fights for top ring events. The Roberto Duran–Boom Boom Mancini and Tim Witherspoon–Larry Holmes fights brought in PPV prices of $15 to $17.50 per set. Arum was able to put together a Hagler–Duran fight in November 1983 that grossed $2.6 million from PPV.

In Cincinnati sex would also be available on a pay-per-view (or pay-per-peek) basis. The Playboy Channel was offered to Cincinnati subscribers of the Warner Amex system there for $5 a night.

In mid-1983, ABC and Cox Cable said their research told them that there was a market for a national PPV network, and they planned a joint venture to test the idea further.

The June 9, 1983, issue of *Daily Variety* reported that a company called FirsTicket had run a seven-month experiment showing movies and concerts to cable subscribers in San Diego and Santa Barbara, and was very encouraged by the results. A confidential HBO memo on pay-per-view surfaced in the November 5, 1984, issue of *Broadcasting* magazine saying that HBO must get into PPV before it was too late. The memo noted that HBO faced an enormous threat from home video and that one of the solutions was to get into pay-per-view as soon as possible. The report went on to make several other telling points. The studios release their movies to the videocassette market at least six months before pay cable TV gets those movies, "so HBO can no longer claim to bring theatrical motion pictures into the consumer home first."

It also noted that PPV hadn't been profitable for a time because there wasn't enough equipment available. However, this was changing, since more and more equipment was being built and more cable operators were installing it. The HBO report predicted that HBO would lose a lot of its customers to PPV unless HBO got into the PPV business quickly.

The focus of the report was on how HBO could sew up the PPV market the way it had almost sewn up the pay TV market.

One of the suggested ways was getting the exclusive rights to movies. Another recommended approach was to control cable channels and freeze out competitors. Time Inc., which owns American Television and Communications (ATC), the second-largest cable system operator in the country, was in a position to do that.

(Not everyone agreed that PPV was going to be the biggest thing to come along in television since the cathode tube. Michael Fuchs still thinks pay-per-view might be premature, and besides, he and HBO handled the threat of home video another way, as we have seen—by getting into home video themselves.)

The report concluded that cable operators were learning to deal with the Hollywood studios directly so as to get movies months before HBO offered them. That could be as disastrous for HBO as having people steal HBO programs without paying for them, and that brings us to pirating.

"It is, it is a glorious thing to be a Pirate King," goes the lyric in Gilbert and Sullivan's *Pirates of Penzance*. Alan Caito of Fairfax, Virginia, agrees with Gilbert and Sullivan, but Joe Collins of HBO in Manhattan would like Caito in jail.

Caito shelled out $5,800 for a satellite receiving dish for his backyard and can now pull in seventy TV channels from a variety of satellites. What is more, he doesn't pay anybody a cent for his fun. Because of Alan Caito—and millions of others like him—HBO is losing millions of dollars every year.

The cable industry has always been plagued by "signal pirates," people who electronically steal cable programming that they should be buying from the local cable operator. Industry analyst Paul Kagan estimated that 3.14 million households steal basic service—that is, about 10 percent of the total number of cable subscribers. There are no accurate counts of how many people steal the premium pay-cable channels, although HBO figures there are at least five hundred thousand homes with backyard receiving dishes.[1]

In its June 6, 1984, edition, *USA Today* quoted Ray Conley, owner of Movie Systems, Inc., as saying that in some cities there is one pirate for every paying customer. The National Cable TV Asso-

---

[1] Of course, there are those wags who note that the local cable operator has been stealing the programming from the networks and the local TV stations without paying for it. That was the crux of the copyright fight that went on for years in and out of Congress.

ciation estimates that the theft of programs cost the cable industry $165 million in 1981; $240 million in 1982; $350 million in 1983; and more than $600 million in 1985.

All of this led some cable program services including HBO to start scrambling their signals in 1986. Scrambling made one satellite dish owner—he called himself "Captain Midnight"—in Florida so mad that he transmitted a power signal to the satellite carrying HBO and blanked out HBO for several minutes. (When he was caught by the FBI, several months later, he was fined $5,000.)

Experts agree that while scrambling the satellite signal will solve the theft problem for a while, somebody will soon be in the business of selling private descramblers to dish owners. It is like using radar to trap speeders. The cops use radar and the private citizen buys a radar detector. Technology keeps on escalating. Still, scrambling the signal so hundreds of thousands of backyard TV Receive Only (TVRO) dish owners can no longer get their television free seemed the only answer to HBO, which has scrambled its picture, and to Showtime/The Movie Channel, which hasn't scrambled its picture at this writing.

However, as with many things in America, this issue has more than just an economic side to it. There is also a political side, which is an aspect to which cable programmers have not always been sensitive in the past. The hundreds of thousands of TVRO owners and their suppliers are voters, and many are vehement about their rights as free American citizens to steal HBO from the sky. Piracy is an issue, in other words, that isn't going to go away soon.

Chapter 20

Merger Mania

"If your biggest customer is becoming bigger than you, you have to go into his business." That was the view of Paul Kagan, and it was shared by almost everybody in the movie studio–pay cable TV game.

At the beginning of 1983, Paramount, Universal, and Warner Brothers all decided to make another run at getting into cable TV. The studios announced they were joining with Warner-Amex, owner of The Movie Channel, and with Viacom, owner of Showtime, to create a joint pay cable TV service: Showtime/The Movie Channel. Only a few weeks before, HBO had gotten into the movie studio business through its partnership with CBS and Columbia Pictures to form Tri-Star Studios.

Naturally, the Justice Department quickly announced that it was going to take a look-see into both the Tri-Star deal and the Showtime/The Movie Channel deal. The movie studio lawyers thought the Showtime/The Movie Channel deal would pass Justice, since it was different from the previous Premiere combination that the Justice Department shot down.

First, unlike the Premiere deal, with Showtime/The Movie Channel the movie studios were not going to reserve their movies exclusively for pay cable service. Second, this was a Justice Department under the Reagan administration, not the Carter administration. The studio lawyers figured it would be much more pro-business.

Jefferson Graham, then a writer for *The Hollywood Reporter*, said that the reason the studios were pushing to take over the two pay TV channels could be found in their 1983 schedule of movie releases. Of the one hundred–plus films coming out of the major studios in 1983, HBO had the exclusive rights to 60 percent of them, thus virtually freezing out Showtime and The Movie Channel. HBO had a stranglehold on Hollywood production and was not shy about using its power to control the marketplace. Oddly, as with the Premiere case we reviewed earlier, the federal government avoided examining whether or not HBO was getting so big as to impede competition—something that the studios have thought all along.

A look at how pay cable affected the overall economics of Hollywood is revealing. For example, even though MCA had a great year—in fact, they had a record-breaking year in 1982, making $469 million in sales in the first nine months—studio executives looked anxiously to the future. As they saw it, five years before, 80 percent of a film's income came from theater showings. However, if the trends evident in 1983 kept going, that number would drop to 50 percent by 1990; much of the other half would be coming from cable TV and home videocassettes.

Even worse, the studios had a fifty-fifty split with theater owners on what is taken in at the box office. With cable TV, however, the studios usually get 20 percent. In addition, Home Box Office is such a big customer, it can almost dictate what it will pay the studios. In fact, most movie production budgets are predicated on a sale to a pay cable programmer, and Home Box Office is the biggest of these.

Add to this the other problems the studios were having—rising costs, militant labor unions, and $1 billion a year lost to pirates who duplicated and sold illegal copies of their films—and it was clear why the studios wanted a piece of cable and home video.

Meanwhile, without awaiting the final approval of the Justice Department, the new HBO-Columbia-CBS company, Tri-Star, announced one of the most ambitious production schedules of any new studio in the memory of Hollywood. Tri-Star planned to produce eight movies in the first two months of operations. The ones slated included *The Natural* with Robert Redford; *The Muppets Take Manhattan*; *Songwriter* with Willie Nelson and Kris Kristofferson; Robert Benton's *The Texas Picture* (released as *Places in the Heart*); *Jimgrim*; and *La Cage Aux Folles, USA*.

As this good news came out for HBO and its partners, some bad news came out for the Showtime/The Movie Channel partners. In May 1983, the Justice Department staff decided to block the merger. At about the same time, Robert Lindsay, the *New York*

Times West Coast bureau chief, did an analysis of what HBO had done to Hollywood. He said it had become the dominant force in Hollywood and that the major studios had basically lost control of the movie business.

Of course, movie studios themselves were not what they used to be. In the so-called golden days of the big moguls, such as Louis B. Mayer, Sam Goldwyn, and the Warner Brothers, a studio was a coordinated, single operating unit. Today, a movie studio is usually a collection of soundstages, specialized departments (technicians, costumes, props, etc.), and administrative units. Often the studio simply rents those components to independent producers as they are needed. Stars are independent, and scripts are written by freelance screenwriters. Today's studio is sort of a Hertz-rent-a-movie-studio.

Some of the most powerful ways in which today's movie studios have a grip on movie production in addition to having the physical, technical, and administrative plant is the studio's firm connections to the financing and distribution of movies. The average independent producer often doesn't have the money or the banking connections to finance his film. Consequently, he has to work out a deal with the rich studio either to finance him or to connect him with a banker who will. When HBO came on the scene willing to put up 50 percent of the cost of a movie before the first foot of film was shot, that made it easy for the producer to get bank backing without involving the studio. This angered the studios because they were cut out of the deal and hence didn't have as much control over the finished movie.

However, the studios still had a tight rein on distribution to movie theaters. Back in the forties the U.S. Justice Department had made studios give up their ownership of movie theaters, but the studios had simply spun off their movie theater properties; they kept a hand in by staying close to the new theater owners, many of whom were friends or colleagues of the studio executives. Both needed each other. What good is a movie theater without movies or a movie without a place to show it?

This closeness to the movie theater distribution network was so tight that the studios commonly engaged in blind bidding and block booking. In blind bidding, theater owners are forced to bid for the rental of movies that they haven't seen or maybe haven't even heard about. In block booking, theater owners were forced to take a block of movies together instead of selecting them individually. In that way, the studio could group some of its cinematic dogs along with its blockbusters so that all the movies would make money.

Here again HBO's appearance on the scene was a lot like a belly dancer at an Anglican religious service. In addition to providing money up front for independent producers, HBO also provided an audience of millions of subscribers for that movie. Cable TV had become one of the major systems of distributing movies, and of course, cable was dominated by HBO.

Robert Lindsay said in *The New York Times* that HBO's rise to power over pay TV while the movie studio executives seemed to be daydreaming over long lunches was one of the biggest success stories in American business. As he saw it, the movie studio people had never been as frightened of the future as they were in the middle of 1983, with HBO dominant and the Justice Department blocking the studios from getting into cable TV.

The Showtime people met with the Justice Department to look for a compromise or reconsideration that would save their deal, but it was no go. It seemed that no matter what Showtime/The Movie Channel and its movie studio partners promised, they couldn't remove what the Justice Department found as the fatal flaw in the whole arrangement: size. The proposed combination was just too big and had the potential of being too powerful. (Again, the Justice Department ignored the fact that HBO was the overwhelming giant in the pay television marketplace.)

A lot of people didn't seem to understand that HBO was already bigger than a couple of the movie studios combined and was pumping out more movies than any of the studios in Hollywood. A multifaceted entertainment company, it rivaled anything going.

Part of the problem was that Hollywood movie people are not, by nature, giant risk-takers. They like sure things. They feel comfortable in a combination with two pay cable services, three studios, and two experienced entertainment conglomerates. The individual risks are small and the profit potential large. The message to Hollywood was clear: Small is beautiful, large is illegal.

Jerry Levin told everybody what HBO had become, but not everybody was listening closely:

> HBO is now the largest purchaser of motion picture rights in the world. But it is not simply a pay TV network anymore. It's kind of a new style merchant bank.

After Justice had said no twice to the Showtime/The Movie Channel deal, the concept of that merger was rethought. The new concept was a Showtime/The Movie Channel/Warner Brothers Studios combo. It would be smaller than the original plan, but it meant that each of the partners had to put up more money and take a bigger risk. Finally, in August 1983, the Justice Department okayed this new arrangement.

Viacom got 50 percent of the new company, with 31 percent going to Warner Communications and 19 percent going to Warner-Amex Cable. Because Warner's The Movie Channel had only 2.5 million subscribers, as opposed to Showtime's 4.5 million subscribers, Warner paid Viacom $40 million to equalize the difference. In addition, Viacom got $5 million a year for six years to act as a "consultant" to Warner Communications, Inc. With that and the $40 million paid out front by Warner, Viacom got a total of $70 million.

This was a good deal for Viacom, since it bought out 50 percent of Showtime from Group W for $75 million back in mid-1982. Viacom ended up paying $5 million for 50 percent of The Movie Channel, and it was agreed that Viacom's Showtime people would be running the new combination service with Showtime's Mike Weinblatt in charge.

Probably even better off was American Express, which, because of its partial ownership in Warner-Amex Cable, ended up with 9.5 percent of the 19 percent Warner-Amex had, and it hadn't cost American Express a dime more.

A week later, *Variety*'s Tom Girard reported that the Justice Department had approved the $400 million Tri-Star Pictures joint venture of CBS, Columbia, and HBO. As 1983 ended, the studios had made some progress, but HBO was still the King Kong of Hollywood.

Chapter 21

Programming, Marketing, and Ratings

The key to pay cable profits is programming, and although HBO was generating more and more programs of its own, what Hollywood produced continued to be a vital ingredient of HBO programming.

The studios were having their ups and downs. For Universal, for instance, 1982 had been a good year, because of the smashing success of *E.T.* But 1983 turned out bad, thanks to a bunch of so-so movies such as *The Pirates of Penzance, Jaws 3-D, Rumble Fish,* and *Doctor Detroit.*

The Universal Studios situation was typical of the love-hate relationship between Hollywood and HBO. HBO was very unpopular at Universal, but the studio had to peddle its dogs to somebody. And HBO was their best customer. Meanwhile, HBO was trying to wriggle free of movie dependence on Hollywood by doing its own original productions and joint production deals, with mixed results. A sample of its projects: *Flashpoint,* the story of two border patrolmen who uncover a secret plot involving the FBI; *Heaven Help Us,* with Donald Sutherland; a return of *The Fraggles,* done by Jim Henson of Muppets fame; an English suspense movie starring George Segal, *The Cold Room; To Catch a King,* Robert Wagner and Teri Garr in a romantic adventure during World War II; and *Robin Hood and the Sorcerer.*

In addition, there would be the Everly Brothers in a reunion concert; a live telecast of the Ray "Boom Boom" Mancini–Bobby Chacon lightweight championship boxing match from Reno;

106 Inside HBO

HBO's first miniseries, *All the Rivers Run;* and *Growing Up Stoned,* another in a series of documentaries by Ann Hassett, at DBA Productions.

Sports and boxing in particular had been good for HBO, but early 1984 saw a troubling debate in the media and sports circles about the Sugar Ray Leonard fight against Kevin Howard. The fight was an HBO exclusive, and it meant a lot to the program schedule.

Sugar Ray Leonard had recently undergone an eye operation, and a lot of people were raising questions about whether he ought to fight or not. (People were particularly sensitive because of the recent tragedies, such as the Korean fighter Kim who had died as the result of a boxing match.) Seth Abraham, HBO sports veep, said that Sugar Ray Leonard was a sensible guy and financially secure. He wouldn't fight if it might endanger his eyes.

Michael Fuchs said his objective for HBO was to offer subscribers a wide range of programming they couldn't see anywhere else. He remained high on movies, with the emphasis on the commercial rather than the artistic—"We want to prove we can make commercially successful movies. The rarefied prestige movies are a luxury we cannot afford."

In response, television critic Tom Shales of *The Washington Post* said that HBO made lousy movies. In his February 11, 1984, column Shales said that no TV network executive should lose any sleep about competition from HBO. In his opinion HBO could make movies just as mediocre as ABC, CBS, or NBC did. He called *To Catch a King* a "laughably inept period thriller" that was unbelievably bad.

Some HBO executives may have taken that column to heart. A few weeks later HBO announced that it was dropping its plans for some of its own movie projects so that it could use the money for buying Hollywood studio films instead. (An unidentified company spokesman told *New York* magazine in the April 2 issue that HBO's own films "have not always been well received." "We're just creatively redistributing our money [from HBO films to Hollywood studio films].")

To get some idea of what HBO's money means to the entertainment industry in general and Hollywood in particular, consider this: Pay television forked over more than $800 million in 1983 for programs, and almost $600 million of that was for movie rights. And, of all those lovely dollars going into producers' pockets from the pay TV industry, half of them came from HBO.

Another concern at that time was Cinemax, HBO's younger sibling. HBO executives were focusing on two major goals in early 1984: (1) how to beef up Cinemax as competition to Showtime/The Movie Channel and to make Cinemax the #2 pay cable channel; and (2) how to find other areas of subscriber growth, since it seemed clear to HBO execs that cable subscriber growth was going to be slow in the future.

On the second goal, HBO executives were looking hard at other forms of distribution of programs besides cable. None looked really appealing except perhaps the exploding home video market, and there HBO had to tiptoe lightly so as not to upset its cable affiliates. Even so, HBO was protecting its position by edging into home video as quietly and as quickly as it could.

On boosting Cinemax, HBO executives were beginning to realize that, in spite of all the thumping they had done in the past about exclusivity, that wasn't what really turned subscribers on. What turned them on was something new, something that hadn't been seen before, something no one else could show them. Of course, exclusive movies could meet part of that consumer hunger, but it could also be met by pay TV "premieres" or "first time to appear on pay TV" programs such as concerts and other specials. So Home Box Office began creating a new Cinemax, one that would be less movie-oriented and would look more like one of the three commercial networks.

The A.C. Nielsen report for 1983 revealed that the number of HBO subscribers was up, but that they were watching less. Nielsen also said that HBO had 13.5 million subscribers and had been adding them at the rate of 2.5 million a year for two years. However, the number of prime-time viewers among those subscribers had slipped from 11.7 percent in January of 1982 to 10.5 percent in January 1983 to 9.4 percent in January 1984.

HBO execs have always maintained that Nielsen ratings are not as important to pay TV as they are to the networks, but even HBO couldn't ignore the ratings entirely. To counter some of the negative news from Nielsen about slipping HBO ratings, HBO decided to start releasing its own reports starting in March.

The decision to release this information was painful, because like many corporate executives, HBO's executives loved to keep anything secret that might reflect poorly on their performance. But the pressure was on.

(The trade newspaper *Multichannel News* put its finger on it precisely in its March 5, 1984, issue when it revealed that Wall Street investment gurus had lowered the rating of Time Inc. stock because HBO's audience was slipping. Remember that Time Inc. was trying to keep its stock prices up to discourage takeovers. Remember also the new Time Inc. focus as expressed a few weeks later at the prestigious business executives' organization based in New York, The Conference Board, by video executive vice president Nick Nicholas. "The first principal objective of our mission is to make Time Inc. a substantially more attractive investment," he said.)

Part of the explanation for HBO's falling ratings in 1984 was that there were more competitors vying for the viewer's time than there were in 1982. Another reason was that there weren't a lot of good movies around. Since 70 percent of HBO's programming was still movies, that hurt. Another problem was that the cable operators were all of a sudden giving the cities that had granted franchises a large dose of economic reality and were cutting back on their construction plans. So, for all these reasons—more competition, shortage of attractive programs, and a decline in cable system construction—there were fewer people watching HBO.

Some example of what was happening and what would continue to happen comes from the experience of communities such as Baltimore, Milwaukee, Dallas, Pittsburgh, Washington, D.C., and Maryland's Montgomery County. While this is getting a little ahead of our story, it might add perspective to what was happening.

The city officials in Baltimore, for example, were stunned that the leading contender for their franchise, Cox Cable, said, "include me out." In another part of Maryland (the lucrative Montgomery County suburb of Washington, D.C.) the United-Tribune cable company said it couldn't make a profit under the franchise terms; in October of 1985, it simply stopped building the system. By June of 1986, United-Tribune and Montgomery County were in a big lawsuit over the issue. Finally, in November of 1986, former Warner-Amex executive Gus Hauser bought out the United-Tribune Montgomery County franchise after making the county politicians give him big concessions from the original franchise contract.

Drew Lewis, formerly Reagan's secretary of transportation, had left government and become head of Warner-Amex Cable, where he would stay until 1986. He tried to renegotiate the Warner-Amex cable franchises in Milwaukee and Dallas and quit the Pittsburgh franchise deal entirely. Said Lewis, "We just promised too much, and now we find out that to break even, we can't live up to those promises."

The city officials in Washington, D.C., granted the cable franchise to a company run by a buddy of the mayor who came back time and time again to ask for better terms for his franchise. Each time he has gotten them—the most recent time being September of 1985. As of the end of 1987, four years after the franchise for Washington, D.C., was granted, there still has not been a single foot of cable laid.

All of this had both a direct and an indirect effect on pay TV and Home Box Office. With fewer systems being built as fast as had been predicted, there were fewer new customers signing up to compensate for those who were disconnecting every year. Besides, fewer people taking cable were buying into premium pay services such as HBO and Showtime. They were using their VCRs instead.

A special report in the March 4, 1984, *New York Times* said that the number of subscribers buying pay cable TV services was down. One survey showed that the average cable TV household subscribed to 1.3 pay services instead of the two that the industry had hoped would be the average. Even more troubling was the fact that 40 percent of all pay subscribers canceled their subscription every year.

At HBO programming, they were hyping Cinemax for all they were worth. A big move was the decision to put *Breathless* and *Tootsie* on Cinemax before or at the same time as they would be shown on HBO. In addition, the plan was to put more and more films on Cinemax to cut down on the repeats that drive subscribers bonkers.

In addition, there would be "new, innovative" comedy, music videos, drama, and such "adult" programs as *Eros America; Mr. & Mrs. Nude California; The Happy Hooker Goes Hollywood; Intimate Moments; Young Lady Chatterly; Scandals;* and *Sex Machine.* All of this was scheduled, as the Cinemax copywriters phrased it, "with sensitivity and discretion." Bridget Potter, HBO's key programming maven, had instructed her staff that the Cinemax programs should be "spicy, but not obscene."

On the comedy side of HBO programming, one of the most popular of the new comedians promoted by Home Box was Rich Hall, the inventor of "Sniglets," new words that Rich made up to describe things in life that the existing words in the dictionary have missed so far. Aside from his success on HBO, his first *Sniglets* book hit the best-seller list before 1984 was over. By 1987, there were Sniglets calendars, paste-on notes, and the full array of promotional items.

One of the programming decisions that had to be made with some sensitivity in the late spring and early summer of 1984 was about one of HBO's major made-for-pay-TV movies, *Sakharov*. The movie was ready for release in September, but real-life events in the saga of the Russian dissident and his wife had jumped ahead of that schedule.

There was serious discussion in the executive suites at HBO about whether or not *Sakharov* should be released earlier to take advantage of the hunger strike that Sakharov had begun May 2 to force the Kremlin to let his wife go abroad for medical treatment.

Some said it would be wrong to exploit the situation, and Michael Fuchs was sensitive to being perceived as commercializing the circumstances. On the other hand, Sakharov's stepchildren, who lived in the United States, and several human rights organizations thought that the immediate showing of the movie would be a boost to public awareness and to Sakharov's cause.

The decision was eventually made to keep to the regular mid-September scheduled release but to have an early special showing on June 20, plus certain private showings. After the June 20 showing of *Sakharov*, David Crook of the *Los Angeles Times* did an article describing the trouble he had getting any kind of information out of Judy Torello and her media-relations staff on the kind of reception *Sakharov* got. He actually wanted to write a story about *Sakharov* and the audience it had garnered, but he ended up writing a story about how HBO people frustrated his efforts to write a story.

Crook said that he had called both the Los Angeles media-relations office and the New York media-relations office of HBO but he had simply gotten the royal runaround.

At a time when HBO's image could have used a little boost, there was nobody there to do it.

Chapter 22

HBO vs. Showtime

Nobody likes to lose, especially a tough-fought, highly competitive game with millions of dollars at stake. That is as good a description as any for the game in pay television in 1983, with the closest rivals being Showtime and Home Box Office. Most of the time, HBO was the winner because it was older, richer, and bigger—more than twice as big as Showtime, in fact.

Still, Showtime had its share of triumphs. Two of the biggest ones in the history of the rivalry happened at the end of 1983. First, there was the Paramount/Showtime deal that gave Showtime the exclusive rights to Paramount movies for five years; and then there was the transfer of 784,000 Spotlight subscribers to Showtime instead of HBO.

This wasn't the only thing that really got the goat of Fuchs, Biondi, Cox, Scheffer, Abraham, and all the other HBO execs. What really hurt was that nobody felt sorry for HBO. Everybody—HBO's friends and enemies alike—thought the Showtime-Paramount deal was good for the industry.

Sharon Rosenthal's assessment of the deal in the March 3, 1984, issue of *TV Guide* was that the deal was the best thing to happen since the coaxial cable made national TV possible. The money that Showtime has committed to the deal, she said, makes this a major act of war against HBO, which had chopped up Hollywood moguls badly. She quoted former Time Inc. executive Austin Furst as characterizing the way HBO did business as the same way General George S. Patton and his Third Armored Division waged war.

One of the reasons many people applauded the deal was that it brought competition back into the pay TV marketplace. Bob

Klingensmith at Paramount expressed the feelings of many people when he said that HBO was on its way to owning all the rights to all the movies made in Hollywood and, in the process, destroying the movie industry. *Broadcasting* magazine, a longtime bible of the electronics business, said in its January 2, 1984, issue that the industry executives loved the deal because it gave HBO its comeuppance and increased competition in a marketplace too long dominated by HBO.

The official HBO response was, of course, that it didn't matter; but no one believed that. Everyone knew the deal would hurt HBO. For example, HBO would not be able to show four of the top ten movies of 1983, *Flashdance, Stayin' Alive, Trading Places,* and *48 Hours.* And the deal gave Showtime/The Movie Channel exclusive rights to all the Paramount productions—probably about seventy-five movies—for the next five years at a cost of around $500 million. That may seem like a lot of money, but it will be cheap in the long run if the gamble pays off.

Beyond these movies, the Paramount deal also gave Showtime *The Dead Zone, Cheech and Chong—Still Smoking, The Lords of Discipline, Uncommon Valor, Testament, The Keep, Footloose, Star Trek III,* and *Racing with the Moon.* Not included in the exclusive list but available to Showtime as well was *Indiana Jones and the Temple of Doom.*

HBO, which has always crowed about exclusivity, made exclusive deals with a number of other studios, such as Columbia, Orion, Silver Screen, and its CBS-Columbia-HBO joint venture, Tri-Star.

Rich Frank, president of the Paramount Television Group, doesn't like HBO, so naturally he loved the deal: "We think it's a fantastic deal. We are getting more [money] than we thought we could."

Writing in the January 8, 1984, *New York Times,* David Hajdu saw that what was happening with Showtime/The Movie Channel was one of the two most significant things to take place in cable for a long time. The other was the deal that gave the subscribers of the Spotlight channel, which was going out of business, to Showtime.

Spotlight had been started by the Times-Mirror Company, which owned a lot of cable systems. Long after HBO had begun by being a broker of movies between the studios and cable operators, Times-Mirror woke up to the fact that cable operators didn't really need a broker, that they could deal directly with the studios. However, they didn't deal directly, and by the time Times-Mirror launched Spotlight, a cable-operator-owned program service dealing directly with the studios, it was probably too late. By that time too many movies were sewn up by HBO, Showtime, or The Movie Channel. After giving it a valiant try, Spotlight decided to fold its tent and transfer its subscribers to some other pay television program service. Naturally, both HBO and Showtime wanted the Spotlight subscribers, but Showtime won.

Getting the 784,000 Spotlight subscribers gave Showtime/The Movie Channel a total of about 8 million subscribers, making it a strong number-two competitor to HBO. Mike Weinblatt, president of Showtime, was ecstatic. Some industry observers were betting that Showtime also had a good run at an exclusive deal with Sid Sheinberg at Universal because HBO had made its pitch to Universal already, and Michael Fuchs and his sidekick Steve Scheffer had returned to New York. (The theory was that they wouldn't have left Hollywood if they had been close to making a deal.) Besides, Sheinberg hated Fuchs and would have liked nothing better than to freeze him out. According to those in the know, after all, it was Sid Sheinberg who labeled HBO "The Crocodile that Ate Hollywood."

Well, there is a weakness in both theories. In the first instance, leaving town was just a bargaining ploy for Fuchs and Scheffer. New York was only a phone call or a five-hour flight away. By leaving town, they were showing that they weren't very worried and weren't that eager for the Universal deal. They were playing it cool.

What most observers didn't know was that Fuchs had canceled plans to attend an important staff outing of several hundred HBO people in Jamaica so that he could be available to negotiate a deal with Universal. (Staff outings, when the company flies people to exotic places for a week of fun, are a big deal at HBO. Fuchs's missing one of those is unusual.)

As for Sid's hating Michael, that was true, but he would not have let his feelings get in the way of business. In fact, Sid is the one who initially went to HBO to suggest talking over an exclusive deal. They were still negotiating, hatred or none. The commonality was money.

Hollywood Reporter writer Jefferson Graham commented on

the irony of a HBO/Universal deal when he wrote for the January 9, 1984, issue. He quoted one pay TV watcher as saying, "It would prove that money talks and bullshit walks."

One of the reasons Sid was negotiating with HBO was that Universal had not been putting out the big movies of late. Sure, it had the biggest money-maker of all time, *E.T.*, and that had gotten Frank Price to come over from Columbia as head of production. Unfortunately, the rest of Universal's movies were so-so films, such as *Doctor Detroit, Nightmares, Jaws 3-D, Psycho II,* and *Sting II.* Universal needed a big-money deal and HBO was a prime prospect for such a deal.

In fact, Frank Price and Universal would have tough sledding for the next several years with a bunch of poor pictures, culminating with *Howard the Duck,* which saw him fired in September of 1986.

Even so, it was a sober Michael Fuchs who returned to New York after his tour of Hollywood in search of film deals. He was muttering that the atmosphere had changed, that he was no longer the Prince of Pay. What's more, the specter of home video was also looming much bigger than Fuchs wanted to let on to the outside world. That would weaken his bargaining position with Hollywood studios, all of which now had home video units and were aggressively pushing sales. Remember, the studios were making major movies available to home video *after* they had been in the theaters but *before* they were on pay cable.

Tom Girard quoted Fuchs's concern in a *Daily Variety* story of January 12, 1984. He said that HBO research suggested that 75 percent of all HBO subscribers would have VCRs by 1988, and they could see movies on videocassette months before those movies would appear on HBO. According to Girard, that worried Michael Fuchs.

Fuchs realized that HBO was going to have to fight harder to get the kind of programming that it needed. He was already being whipsawed by his own contradictory public statements on exclusivity. For a long time, he preached the necessity of exclusive contracts for movies, but after the Paramount/Showtime deal, he began to downplay exclusivity. Predictably, Showtime played *up* its exclusive deal with Paramount. For instance, it was going to show *Flashdance* in February, and it promoted it everywhere as being "exclusively on Showtime."

Meanwhile, Michael Fuchs was backpedaling a little more. In

early March, at a West Coast appearance supposedly to promote Cinemax, Fuchs declared exclusivity dead. He said that it was a myth that exclusive movies on pay services helped to build an appetite among subscribers for more than one pay service. Fuchs now contended that nonexclusivity was the new religion and admitted, to the astonishment of some Hollywood observers in attendance, that perhaps HBO had been a little too greedy in the past.

Soon after Fuchs's announcement that the god of exclusivity was dead, HBO made the most expensive pre-buy in its history, for the exclusive (yes, exclusive) rights to *Rambo*, a Sylvester Stallone movie that was released in 1985.

In addition, Sid Sheinberg and Michael Fuchs finally made a deal between HBO and Universal. The amount was $200 million, which sounds high for a nonexclusive deal but it really wasn't. Universal expected to produce about 120 movies during the term of its HBO deal, and that works out to about $1.7 million per picture. If you factor in inflation and some HBO subscriber growth, it comes out to about what HBO traditionally has been paying for films.

The soothing effect of money had glossed over the friction between HBO and Universal. As Universal exec Gene Giaquinto said, "Our relationship with HBO has improved dramatically. There's a peace pipe being passed between us now."

I wonder what was in it.

PART IV

HBO AND TIME INC.'S TIME OF TROUBLES

Chapter 23

HBO and Hollywood Try a Truce

On New Year's Day 1984, *The Washington Post* carried a story by Merrill Brown that summed up how Hollywood had continually shot itself in the foot by rejecting new technology. Brown said they had fought against "talkies"; against radio; against color; against television; against cable TV; against pay TV; and against home video. Then, when the previously reviled technology became popular, the movie studios went nuts trying to play catch-up.

Brown quoted Lee Isgur, a vice president and analyst with Paine Webber Mitchell Hutchins, Inc.: "They gave away the store, and they're trying to get it back. What makes many of the squabbles so rich in irony is the fact that the networks, HBO, and the film studios live increasingly interdependent lives."

Merrill Brown saw Hollywood's bitter resentment of HBO's style and tactics as due to an intense cultural difference between the way business is conducted in Los Angeles and in New York. The cocky, abrasive New York style does not play well in the lower-key, laid-back California culture.

In April, everybody got an interesting insight into the mental set of HBO executives when HBO chairman Frank Biondi spoke to members of the Academy of Television Arts and Sciences at Los Angeles' Century Plaza Hotel. Biondi declared peace with Hollywood, saying that HBO's deliberate six-year war of rhetoric and nerves was over. He said that, since cable subscriptions were slowing down, a new and friendly relationship between HBO and Hollywood was appropriate.

What did the Biondi declaration of peace with Hollywood

really mean? Insiders on both coasts saw it this way: First, it was the only time HBO has admitted in public that it had been deliberately waging conscious war against the studios. In previous public remarks, the idea of such a war had been dismissed as Hollywood paranoia. The universal assessment of HBO in Hollywood had always been that it intentionally was abrasive, unreasonable, and mean-spirited as a part of the New York business personality and as a negotiating tactic. Wear the other person down, frustrate and make your opponent angry, and finally—just to get rid of the aggravation—you will make a better deal than you could have gotten by being nice. This is the way Hollywood saw it all along, but HBO rejected that publicly as whining. Now, the president of the company was saying it had all been planned that way just as Hollywood said.

Second, to have war, peace, or tango, it takes two. HBO wanted peace with the studios because it suited Home Box Office's own purposes. That is, HBO was in financial trouble. It wasn't losing money—far from it—but it was seeing a significant drop in earnings. Just as a lot of cable systems were going back to renegotiate franchises with cities, HBO wanted to renegotiate film licensing contracts with the studios based on the new reality of smaller audiences and lower revenues for HBO.

Third, Biondi talked about how tough things had gotten with lower subscriber growth and the explosion of home video. But the studios didn't give a damn about lower revenues for HBO. The studio had made their deal with HBO. And the studios didn't see home video as a threat; they regarded it as a golden opportunity.

A couple of days after Biondi's speech, *Newsweek* published a story, "HBO vs. Showtime vs. VCR," that put the finger on some of the problems facing HBO and the cable pay TV business. It said that HBO was still on the top, but business was slipping; HBO subscriptions were off 30 percent and Cinemax subscriptions were down 75 percent. HBO still made $125 million in net profit, but VCRs were cutting into its cable income.

In its July 9, 1984, issue, *Business Week* picked up the story about the changing and disappointing fortunes of HBO. The story noted that things had seemed rosy for pay TV just a couple of years ago when subscriptions were booming, but a stunning drop in subscriptions at the beginning of 1984 had everybody scrambling to figure out what happened.

It said that all of this disappointing news had depressed stock market analysts, and when analysts get depressed, so do stock prices. A report about HBO by the securities firm of Donaldson, Lufkin & Jenrette apparently pushed a lot of panic buttons along Wall Street. Time Inc. stock dropped almost 15 percent overnight.

Biondi's early projections of 2 million new subscribers now looked way off. The *Business Week* story reported that HBO had only added three hundred thousand new subscribers in the first quarter of 1984.

HBO executives reacted in variations on the same theme: Nick Nicholas, Jr., head of the Video Group, said it wasn't ruination, but it was sobering. HBO Chairman Frank Biondi said, "[If] your revenues don't grow as anticipated, certainly there is a pressure on margins. We are re-examining all our options."

While the *Newsweek* story laid much of the blame on the exploding growth of the home videocassette recorder, *Business Week* suggested that the theory lacked solid proof. In fact, some evidence pointed to the fact that VCR owners were more, not less, likely to be cable subscribers, too.

Business Week brought up that quote from Nick Nicholas's panel appearance at the National Cable TV Association at Las Vegas' MGM Grand Hotel a few weeks before—one that would haunt him for a long time. He said,

> We've assumed that consumers could have three channels that carried virtually the same movies, and they wouldn't know the difference. . . . I would call it a fraud that we perpetrated on the consumer.

Business Week writers pinpointed the situation when they ticked off the problems still facing cable TV: the competition from VCRs, lack of good marketing skills by cable operators, viewer unhappiness with program repetition, and the shortage of good movies.

Even so, Time Inc. was doing pretty well overall. First-quarter results for Time Inc. in 1984 showed an increase—a good increase. The stock made 67 cents a share in earnings, compared with only 35 cents a share in the first quarter of the year before. Earnings from the video part of Time Inc. were also up, but not nearly so much as had been hoped earlier.

It had been widely discussed in the private meeting rooms of the Time Inc. building that earnings for the entire company would be lumped together so as not to reveal what was expected to be a sharp drop in the growth of video's profits when compared to previous years. The plan was to confuse the folks with vague and complicated figures.

The trouble is that sometimes it seems that the Time Inc. and HBO executives are the ones who get confused.

Chapter 24

The Warnings HBO Ignored

Bonanza or bust? In 1984–85, that was the question about cable TV in the urban areas of America. Some people saw the millions of unwired homes in the urban areas as a gold mine. Others said it would be a bust, because the people were poorer and the construction costs were higher.

There is one thing that was very clear then and is even clearer now. The rush to build cable systems in the big cities is over. Even where franchises have been granted, the building will take lots of time and lots of money.

With all the hoopla about cable, the urban areas of most major cities are still not wired for cable. As we said, most of New York and Los Angeles are not wired, and none of Chicago, St. Louis, Baltimore, or Washington, D.C., is wired.

Laura Landro, writing in *The Wall Street Journal,* made the point that even in those cities where franchises have been granted, the changing economics of cable have sharply slowed or halted construction.

The massive slowdown in urban cable system construction was well known, well documented, and well reported throughout the cable industry and the business press as early as the beginning of 1983. Even so, HBO executives claim to have been caught completely by surprise by the sharp drop in new subscriber growth one year later.

Most HBO executives said the same thing that HBO executive Bill Hooks said to me: "It was as if a curtain came down on us New Year's Day [of 1984]."

124 *Inside HBO*

I found that hard to believe then, and I find it even harder to believe now. The warning signs were everywhere back then. What's more, HBO's own research department was clipping the news from reports and journals and putting it on every HBO executive's desk every day. Employees didn't even have to go downstairs to the basement of the Time-Life Building and pay money at the newsstand; the news was analyzed, clipped, and plunked down on their desks every morning by the time of the first coffee break. There were signals everywhere.

For example, a signal came from the regular FCC report on cable industry finances. Being slow, the FCC is usually a year or more late, so it wasn't until the beginning of 1983 that the feds came out with the figures for 1981. Associated Press reporter Norman Black wrote the story in late March.

His story reported that cable TV profits had nosedived 76 percent in 1981, even though there had been a sharp rise in gross income. Net profit dropped from $168.1 million in 1980 to $40 million in 1981. The difference was that it was costing more to build new systems. While HBO was not in the hardware business of building new systems, when the cable operators—who were in that business—suffered, it was a clear signal to HBO.

There were more warning signs of subscriber growth slowdown in the spring of 1983 when Paul Kagan Associates released its report on pay cable growth figures for 1982. Yes, they were up, but a closer look revealed that the rate of growth was slowing.

A. C. Nielsen was reporting the same thing about the viewership for WTBS, Cable News Network, ESPN, and CBN Cable Network. In short, there was an overall drop in the number of people looking at cable during prime time.

Nielsen wasn't the only messenger with bad news. The trade newspaper *Multichannel News* reported on July 25, 1983, that subscribers were becoming more and more bored and unhappy with what was being fed them by pay television. Commercial television was bad and boring and cable television was simply a bland echo. It was not better television; it was just more television.

A survey of sixty-six thousand homes by NPD Electronic Media Tracking Services reported that the longer a subscriber had cable, the less likely he or she was to subscribe to pay cable services such as HBO or Showtime. The reasons most often given for dropping premium cable services were that those services weren't worth the money, movies were repeated too many times, and the quality of the programs was lousy. Sixty-four percent of those responding were ex-HBO subscribers.

All this churning and these complaints were nothing new to HBO executives; they'd heard it all many times before. But what they seemed unable to understand was that things were going to get a lot worse. Unable to cope with the causes of subscriber disaffection, the HBO people continued with business as usual.

Some HBO executives did understand that there was a problem, but they didn't seem to know how to solve it. For example, at the 1984 Cable Television Administration and Marketing Society (CTAM) conference in San Diego, John Billock, senior veep of marketing at HBO, spoke about what the industry had to do when it became "mature." (He meant when cable became a mature business in the sense that cities were all wired and there were no more vast pools of unconnectable subscribers available.) He warned that the industry had better start worrying about what it was going to do when it really had to deliver on all the glorious promises it had made.

Billock set up a panel discussion to explore the question of whether or not the churn rates were largely due to cable's being oversold or underdelivered. The idea of the panel was good, but it missed the point, which was that the buying public was getting shafted either way.

All during 1982 and 1983, the signs were coming in to the HBO floors of the Time-Life Building that business was dropping, dropping, dropping, but a lot of HBO executives seemed to ignore them. In early November 1983, Time Inc. and HBO people got a jolt when the investment firm of L.F. Rothschild, Unterberg, Towbin took Time Inc. off its recommended buy list. The investment house said that the failure of *TV-Cable Week* and dropping earnings in the Video Group made the stock less attractive. Alan Gottesman, a vice president of the investment firm, said that the cable business was not growing as fast as it once had and that fewer subscribers would be available in the future because of the slowdown of cable system construction.

In Gottesman's opinion this problem, linked with the continued problem of churning and the need to try to win back previous subscribers—a tougher sell than getting a new subscriber—all affected the future profitability of Time Inc. stock.

Another alert came from a study by Television Audience Assessment, Inc., which sampled television, both cable and noncable, from three thousand homes in Connecticut and Missouri. The results were that cable subscribers did not find cable television any more enjoyable than commercial network television. And those that didn't subscribe to cable said they didn't think it was that much better than over-the-air TV. They said they wouldn't pay extra for it.

Around Thanksgiving, still another study about why subscribers disconnect was finished by Warner-Amex and released to the press. *The New York Times* of November 27, 1983, reported the results in a story by Sandra Salmans. She reported that HBO had a monthly disconnect rate averaging 3.7 percent of its subscribers. Annually that meant almost 40 percent of its 13 million subscribers disconnected from HBO. The arithmetic is simple: for every ten new subscribers signed up each year, it kept only six.

At the Palm Springs management meeting in late 1983, one of the Time Inc. directors, Henry Goodrich, sounded another of those warning signals that HBO people seemed deaf to. He said, "As the video business begins to mature, it isn't going to grow as explosively as it has in the past."

Perhaps HBO executives ignored the warning signals because they thought that the troubles wouldn't affect them. But they were wrong; some of them were out on the street a few months later because of staff cutbacks at HBO.

Chapter 25

The Most Hated and Feared Man in Hollywood

Early in 1984, when Hollywood heard the news from New York, the shudder could be felt all the way from the New York headquarters of Home Box Office to the Polo Lounge of the Beverly Hills Hotel. Michael Fuchs, the Prince of Pay TV, had been made the new president of Home Box Office.

The shudder turned into a Richter Scale 10 earthquake several months later at Halloween, when his boss, Frank "The Golden Boy" Biondi, was suddenly and unceremoniously dumped by Time Inc. and Fuchs was moved up to chief executive officer of HBO. The Prince of Pay had been transformed into the King of Pay.

In Hollywood, the stunned movie folk could hardly talk of anything else at Ma Maison's in the days that followed. Of course, much had already been said about Michael Fuchs. *Esquire* magazine has called him "the most potent, feared and hated man in Hollywood." The August 15, 1983, issue of *The Wall Street Journal* ran a profile of the two things that Hollywood thinks was wrong with pay television. Those two things were HBO and Michael Fuchs.

In the words of one Hollywood producer, "Greed, avarice, and hatefulness—you can't eliminate that from the Michael Fuchs story." Or, as one movie agent put it, "This industry is going to get Michael Fuchs. Just you watch."

It was unlikely that anybody was going to get Michael Fuchs

any time soon. In 1984, at the age of thirty-eight, Fuchs had become the most powerful man in the entertainment industry.

It is hard to tell exactly why so many people hate Michael Fuchs. Is it because of his abrasiveness or his hard-nosed negotiating techniques? Perhaps.

Fuchs himself admits that he's confrontational and aggressive, with a single-minded drive to be a winner. Others say that he is the way he is only because he is backed by Time Inc., which is using him as "our Jew" to fight against "their Jews" in Hollywood. (That sounds dreadfully anti-Semitic, but it happens to be a factor in all of this. It is similar to the observations that Dan Jenkins made in *Semi-Tough*. Jenkins says that the main issue between professional football teams is whether "our niggers can beat their niggers.")

It is well known on both coasts that Fuchs has a very thin skin and that he hates it when people are critical of him. He's noted that America always roots for the underdog until he becomes the upperdog, and then the public turns on him. In some ways he's right, but I think whether America roots for the underdog or the upperdog is partially due to his personality. Michael Fuchs is definitely not what anyone would call lovable.

Fuchs publicly brushes off criticism by saying that he may be Sicilian. He is patently not Sicilian, but what Michael Fuchs is can be divined by his own admission that, while at the William Morris theatrical agency, he steamed open other people's confidential mail so he could get a bargaining advantage over them.

As top man at HBO, Fuchs decides what HBO's millions of subscribers would see. Because of HBO's growing involvement in home video and other entertainment and because of its enormous financial clout in Hollywood, Fuchs can dictate much of what kind of movies America sees, even to those who are not HBO subscribers. Still, that's not really what has irked Hollywood. After all, most of that money ultimately slides west from New York. It was also not that producers objected to HBO's immense success. No. What irked Hollywood then and what still irks Hollywood today is Michael Fuchs himself.

They say he is petty and vindictive. They say he has thrown away the unwritten rules that have made Hollywood work for decades. One studio vice president says that HBO negotiates as if everything is personal instead of business. Studios, he says, are used to bargaining tough with people, and then going out drinking with them. With HBO it is different. If HBO doesn't get what it wants, HBO turns it into a vendetta.

One typical story makes the point. Paramount Pictures produced a Frank Sinatra special with an unofficial agreement to sell it to HBO. Showtime bid higher than HBO, and Paramount decided to sell to it. According to Paramount's video president, Mel Harris, Fuchs called him on the phone and said, "I'll get this out of your hide sometime." And Fuchs did.

Paramount didn't make another movie deal with HBO for quite some time. For a long time, Fuchs wouldn't even talk to the executives at Paramount. He was teaching them a lesson, a lesson he had taught before on many occasions. However, this time his confrontational style backfired.

In December 1983, Paramount signed a deal to sell all pay TV rights to Showtime/The Movie Channel on an exclusive basis for *five* years. The word of this giant deal broke at the Western Cable Show in Anaheim and stunned the pay cable world. In the minds of many people in Hollywood and New York, this blockbuster deal was part of Paramount's revenge on Michael Fuchs and HBO. It was a way of paying Fuchs and HBO back for their boycott of Paramount after the Frank Sinatra special incident of months before.

> "I was lied to," Fuchs told *Esquire* about the Sinatra affair. "I was furious about being lied to. People can say revenge, whatever, but lying is in a separate category in life."

Yet even those who hate Fuchs find it tough to argue with his success or his chutzpah. Even his joining HBO showed a certain flair. Home Box Office tried to recruit Fuchs's boss at the William Morris Agency, and when his boss turned down the offer, Fuchs called HBO and offered his services instead. HBO took him.

His first job was as director of special programming and sports. When he created the shows "On Location" and "Standing Room Only," he brought stars like Robert Klein and Bette Midler to cable TV. He feels largely responsible for giving Robin Williams and Steve Martin their big breaks, because they were virtually unknown before their HBO appearances. A boxing buff, Fuchs has come close to cornering the heavyweight title matches for the HBO market.

But movies are HBO's mainstay, and it is in that area of entertainment that Fuchs has made his greatest impact in creating made-for-pay-TV movies and in expanding the concept of pre-buys that was left him by his predecessor, Austin Furst. Pre-buys can partially or totally shut out the movie studios.

As for the made-for-pay-TV movies financed by HBO, they cost considerably less than regular feature films but more than network TV movies, and are considerably slicker. Audiences seem to like these movies, and Fuchs has been able to attract some big stars to his ventures—Carol Burnett, Elizabeth Taylor, and Jimmy Stewart, among others. More importantly, perhaps, the pay TV movies have brought HBO some valuable intangibles: prestige and even greater bargaining power. No wonder Hollywood hates Fuchs.

If Fuchs feels the same way about his detractors as they do about him, it doesn't show. With his six-figure income, his Manhattan apartment with private gym, his suites at the Beverly Hills Hilton, and his limos, maybe he thinks that living well is the best revenge.

Chapter 26

Frank Biondi's Success and Failure

In *The Graduate,* Dustin Hoffman was told that the wave of the future was plastics, but a bright young person seeking the business fast track today is better advised to go for something a little more high-tech: computers, electronics, cable TV. Nobody knows this better than the baby-faced, curly-haired, genial-looking Frank Biondi, Jr., the man with the driving ambition and the ready smile.

In 1984, at thirty-nine, he culminated six years of work at HBO by leapfrogging over the rest of the HBO executives, including the man who brought him into the company in the first place, Michael Fuchs. Frank Biondi became chairman of the board.

During his short time at HBO, Biondi had seen its monthly subscriber universe grow from slightly over a million to an estimated 15 million. Under Biondi, HBO became the single most profitable operation in the Time Inc. empire.

Biondi was known for pulling off some deals that set Hollywood on its ear. One might even say that these deals forever changed the basic business practices of the movie business. The December 1984 edition of *Esquire* magazine was devoted to "The Best of the New Generation—The Men and Women under Forty Who Are Changing America." Biondi was one of those included in *Esquire*'s list. He is described as a tough negotiator and smart financier who guided HBO during some of its best moneymaking years without putting his own interests ahead of HBO or his HBO colleagues.

Biondi got a bachelor's degree from Princeton and an MBA

from Harvard. Then he set out on a career in investment banking. Deliberately avoiding large corporations, he focused on working for smaller firms, because he felt he could come in closer to the top. Ironically, the companies he worked for were later swallowed up by giants like Prudential-Bache and Shearson/American Express.

He shifted from investment banking into cable television when he joined TelePrompter, the largest cable system conglomerate in the country, but things did not go well for him there. In fact, the only thing Biondi remembers as worthwhile to come out of that association was his meeting Carol Oughton, whom he later married.

Leaving TelePrompter, Biondi decided to go back to his "small is beautiful" philosophy and set up his own financial consulting firm. That led to his involvement with Children's Television Workshop, which was toying with the idea of investing in some cable systems. From there it was but a short step to HBO.

Even before joining HBO, Biondi was astute enough to recognize the company's Achilles heel. In 1976, he told his friend Michael Fuchs, who had just joined HBO as director of special programming, that HBO was only a middleman. He warned Fuchs that HBO was vulnerable because it was between the cable TV system operators, who have a monopoly franchise in their communities, and the movie producers, who create the main source of the most popular public programming, movies.

One of the great dangers as Biondi saw it was that either side, the cable operators or the movie producers, might be smart enough to realize that they didn't really need a middleman. They could deal directly with each other or get into each other's business or both.

This, of course, is exactly what happened, but by then HBO had won a dominant position in the industry. Basically, Biondi's chief contribution to the success of HBO was to change its focus from being only a middleman for movie producers and cable system operators to being a diversified entertainment conglomerate into most facets of television and motion pictures. Wisely, he saw that cable systems were only one way to deliver a movie or other program to the consumer, and that as technology changed, it would hurt the person stuck with a lot of expensive cable systems. It was better to be the owner of *software* instead of a lot of outdated and expensive hardware.

Biondi understood the classic business school case study about the railroad business in America. In the early part of this century, the railroads were rich and successful. When the truck and airplane came along the railroads scorned them, because, said the railroad men, "we are in the railroad business." In time, of course, the airplane and the truck took a major portion of the railroads' business away from them. As every budding MBA is taught, the fundamental mistake that the railroads made was to think they were in the *railroad* business when, in fact, they were in the *transportation* business. The railroad just happened to be the current technology of transportation. They should have used their wealth and power to adapt to the new technology of transportation when it came along.

Following that same reasoning, Biondi taught the executives at HBO that they were not in the cable television business, they were in the entertainment business. They must be prepared to adapt to the changing technology for producing and distributing entertainment to the consumer who pays the bill and must not be locked in with technology that might become archaic.

Naturally, Biondi recognized the fact that for now the cable system was the main distributor of HBO programs. So he had to be sure that the cable system operator was not aggravated or offended. Still, Biondi believed that HBO should be ready to move into other systems of delivering entertainment to the consumer whenever that system became workable.

When Biondi joined HBO, he quickly hit upon a better way of solidifying HBO's business position. Using his own financial training and Time Inc.'s resources, Biondi appealed to the movie studios' greed. He devised a variety of financing plans for funding movies. By controlling vast pools of dollars, HBO could whistle the tune that movie producers would dance to on command. (This was his other major contribution to the success of HBO.)

He began offering to finance the making of films for the chronically money-tight studios. Austin Furst had done the same thing for independent producers with his pre-buy plan. In return, he demanded "exclusivity," the sole right to show the film on pay cable before it was released to commercial TV networks. The "exclusivity" appealed to the viewer and the advance money appealed to the producers. Beyond that, money has time value—that is, money today is worth more than money next year. HBO money up front meant more to a movie producer than money paid after the film was complete, because it was money on which the producer didn't have to pay interest, as explained earlier.

In return, the bait of HBO's money allowed Biondi and crew to demand more from the producers. One of the things HBO demanded was exclusivity, which eventually included home video, theater exhibition, and broadcast television.

The first studio to bite was Columbia. HBO got first option on all Columbia films and exclusivity on any four in return for HBO's production money. The second studio to go for the Biondi concept was Orion, which, in one of those typically complicated Hollywood deals, had left Warner Brothers, which was distributing its films, and was bought out by Filmways. HBO helped Filmways come up with the money to buy Orion and, as part of the deal, got exclusive rights to Orion's films.

But exclusivity was not enough. Back at the ranch, Home Box Office had gone to twenty-four-hour, seven-day service to keep up with the competition. Biondi, by this time executive vice president for planning and administration, again drew on his investment banking background. He helped work out a joint venture with Columbia, CBS, and HBO for the creation of Tri-Star Pictures, a brand-new movie studio that could create a dozen or so movies a year. Then, working with E.F. Hutton, Biondi helped to develop a new public company, Silver Screen Partners, which raised $125 million to produce another dozen films a year for HBO. HBO had suddenly become the world's single largest financier of feature films.

The boys on the executive floor of Time Inc. were impressed, and they showed how *much* they were impressed by jumping Biondi over his friend Michael Fuchs and making Biondi the new president and chief operating officer of HBO.

Sometimes people riding as high as Frank Biondi was at this point get dizzy or don't think as clearly as they might. And for whatever reason, Frank Biondi made a semifatal career mistake.

Naturally, Biondi was still regarded as something of an outsider, but the Time Inc. insiders had offered him the highest accolade of all, admission to the inner circle of executives. Once someone has joined the blood brotherhood of Time Inc., he becomes not a worker for a corporation, but a member of a family. A family that is caring, nurturing, comforting, and above all, loyal. However, these precious traits must flow both ways. But Frank Biondi didn't get that message. He forgot about loyalty.

Frank stunned his Time Inc. family when he announced, one year after being offered admission to the inner sanctum, that he had been down the street talking to the people at Showtime and that they had made him a very attractive offer. He said that he would prefer to stay at HBO, provided his compensation was improved and his authority clearly established over his two rivals for power at HBO, Tony Cox and Michael Fuchs.

Biondi's treachery, as some regarded it, caught Time Inc.'s inner circle by surprise. Biondi was a valuable player, and they needed him. What's more, since they had just made some executive shifts, it wasn't a good time for dramatic changes at the top of HBO.

However, from that moment on, Biondi was finished at Time Inc.—it was just a matter of how and when to let him off.

That was unfortunate, because of all the places Biondi had worked, it was here that he seemed the happiest. At the age of thirty-nine as chairman of the board and chief executive officer of Home Box Office, Frank Biondi had gone far and fast. One reporter said that Biondi had more clout in Hollywood than anyone except George Lucas. At that moment, the opportunities seemed only to have begun.

Yet by Halloween of 1984, just seven months later, it had all changed. *Business Week* magazine told the story of what happened. Time Inc.'s inner circle had moved one of their own, Nick Nicholas, to take charge of the entire Video Group. Within weeks, Nicholas took direct control of HBO and told Biondi to move out of his corner office.

The announcement of Biondi's exit was reported in *The New York Times* and *Washington Post* of October 13, 1984. Both stories reported the surprise within the TV industry, Wall Street, and HBO itself at the dumping of Biondi. Profits at HBO had not been as high as they had been before, but that didn't seem to be the reason for Biondi's ouster. In fact, other Time Inc. executives in the same situation hadn't been dumped. For example, after the $47 million debacle of *TV-Cable Week*, Magazine Group head Kelso Sutton had not only remained in his job, he had gotten a raise.

No, there was something more involved. Two weeks later *Cablevision,* a cable industry trade journal, said that while Biondi's getting dumped from the HBO slot had been expected, there was something unusual about the way Time Inc. handled the Biondi matter. What was unusual was that instead of being shifted to another job inside Time Inc., Frank Biondi was out on the street.

But then, of course, Frank Biondi had been disloyal to the family. Now, he disappeared into a made-up job with Coca-Cola's entertainment division. He would not surface again until mid-1987 and, then, it would be in a big way facing off against HBO.

Chapter 27

A New Hand at the Wheel

Cutting off Biondi's corporate head was not going to solve all the problems that faced HBO at the end of 1984 and into 1985. (Even today, in 1988, not all of them are solved.) At the time Biondi left, Mara Miesnieks, who analyzes leisure-time companies for Smith Barney, Harris, Upham & Company, said that his departure was only a symptom of the fact that HBO was in some trouble.

Basically, HBO had made the strategic mistake of assuming its number of subscribers would keep growing, but of course, subscriber growth was slowing down. As we saw in a previous chapter, this slowdown was clearly predictable, but people at HBO seemed not to be paying attention. They got caught by surprise with lower growth, higher relative costs, and falling profits.

Soon after Biondi left, there began a massive firing at HBO. One of my close sources inside HBO at the time this was happening described the mood as one of "sheer panic." He also said that Biondi was just a scapegoat; the policies that had gotten HBO into trouble were not Biondi's but rather Jerry Levin's and Michael Fuchs's. It was Levin and Fuchs who had made bad strategy and programming decisions.

Biondi was not the last executive to be shot out of the HBO cannon. A few days later, David Meister and Michael Lambert, both senior vice presidents involved in expanding HBO's operations into home videocassettes and pay-per-view among other things, got the axe, too. Before the bloodletting was over, several hundred HBO employees were trundled to the chopping block. It was the typical corporate approach. Most of the guys at the top stay, but those at the bottom are fired. It happens that way in a lot of corporations. As one friend of mine who got fired during Thanksgiving weekend of 1984 at HBO commented, "At least when the *Titanic* went down, the band was playing."

Firing people, however, was not all that had to be done. Some new people had to be brought in to correct the situation.

After the Biondi debacle and Michael Fuchs's move up to chairman and CEO of HBO, Time Inc. cast around quickly for someone else to fill the presidency. They needed someone who could put HBO back on a solid business track. Joe Collins, president of ATC, Time Inc.'s cable system subsidiary, seemed to be that man.

When you see Joseph L. Collins, the nose is the first thing you notice. It's very prominent and wide at the bottom in the middle of what strikes you immediately as a friendly, Irish-looking face that you'd expect to meet over a Bass ale at the Dubliner bar in Washington or Joyce's on the East Side in New York.

One of the main qualities Collins brought with him to his new HBO job was an understanding of cable system operators, which is where both Time Inc. and HBO had had a lot of trouble in the past. Most cable operators viewed HBO people as arrogant, immature New Yorkers trying to tell the cable operators in the rest of America what they must do for the benefit of HBO.

Collins is also regarded as a tough manager, one who knows how to handle day-to-day operations. It was hoped that he would bring HBO the kind of lean, mean, disciplined operation it should have had a long time ago. With a degree in international relations from Brown University and his MBA from Harvard, he went to work for Time Inc. with ATC in 1972, the same year HBO was born.

The trade magazine *Broadcasting* said at the time, "One of the most frequent phrases used by his industry colleagues to describe Collins is 'solid.' "

And solid was something HBO needed.

Chapter 28

Inside the Executive Suites

In 1985, Time Inc. was in trouble. In the first place, contrary to popular belief, it was no longer primarily a magazine publishing company; for several years it had been primarily a video entertainment and cable TV company.

The Video Group was now the main source of net profit for the company. The big bucks were made by ATC, its cable company, and HBO, its video programming company. What was happening to cable TV and to video entertainment programming deeply affected the fortunes of Time Inc.

Wall Street definitely noticed that things weren't going well at Time Inc. They saw several things that would affect Time Inc.'s cash cow, HBO: slowdown in cable construction; continued churning among subscribers; stronger competition; higher expense for HBO's programming and promotion costs to counteract higher competition and churn; and a price war between HBO and Showtime.

About half of Time Inc.'s stock was held by large institutional investors—banks, colleges, pension funds, mutual funds, and trust funds. The prospect of lower earnings made these people nervous.

Paul Kagan had speculated that if HBO didn't grow by its projected 2 million new subscribers in 1984 and Cinemax fell short of its projected 1 million new subscribers, it could cut Time Inc.'s profits for the year by $3.4 million. If that happened, Kagan feared that institutional investor confidence would be shaken badly.

In another move to defend against a takeover, Dick Munro, Time Inc.'s president, got the Time Inc. directors to okay repurchase of 2 million shares of the company's stock. By cutting down on the number of shares on the market, it tended to raise the value of the remaining shares and reduce the number of "strangers" who have stock and might be lured into selling to a corporate raider.

Munro told a *Business Week* reporter for an article featured in the magazine's February 23, 1984, issue (entitled "Humbled And More Cautious, Time Inc. Marches On—After Four Big Failures, It Searches For Safer Investments in Publishing and Video") that Time Inc. planned to exercise more managerial control over its operations. The story described the stunned confusion that filled the executive suite at Time Inc. after the disasters of *TV-Cable Week, The Washington Star,* teletext, and subscription television (STV), which sapped some $207 million from its coffers. The shell-shocked Time Inc. executives were wandering around wondering what had gone wrong.

The *Business Week* article reported that Time Inc. would be looking for some magazines to buy, and quotes Munro as saying, "I think the day of generating only our own ideas is over. Inventing magazines is tough. We're receptive to acquisitions." *Business Week* quoted one observer as saying, "Don't expect miracles anymore at Time Inc. When companies start looking for acquisitions, it is often a sign that they have run out of ideas."

Munro himself attributes Time Inc.'s problems to one core characteristic—sheer arrogance. In an interview with *Business Week,* Munro repeated that phrase over and over again as the root cause of Time Inc.'s and HBO's problems. For example, no market test was done before launching the ill-fated *TV-Cable Week,* because Kelso Sutton and other executives of the Magazine Group didn't think they needed a market test. After all, they thought, they were professionals. They knew what would work.

It was with the same mental set, Munro said, that Time Inc. bought *The Washington Star* in 1978. The paper was failing, but Time Inc. executives thought the nation needed them to own a newspaper in Washington, D.C., and to compete with *The Washington Post,* who owned its magazine rival *Newsweek.* Having bought the *Star,* they proceeded to put a man in charge who had never put out a daily newspaper before.

The story went on to point out that HBO, which had been

yielding a 25 percent profit on its revenues (the Magazine Group only made about half that profit) was facing a maturing marketplace with tougher competition, slower growth, and at least a 33 percent churn rate.

On March 20, 1984, Nick Nicholas gave a confidential speech to the prestigious Conference Board in New York on the search for Time Inc.'s soul. The essence of that speech was that Time Inc. was a corporation beset by uncertainty over what and who it was, as well as where it was going. He revealed that Time Inc. was painfully aware of its own confusion and noted that some financial analysts were as confused about Time's confusion as Time was. "The investment community viewed our planning style either with popeyed admiration or total bafflement," he said. Time had a strong balance sheet, he said, but because the stock market was confused about what the company was, the price of its stock tended to be lower.

Henry Luce had founded a company that was "to serve the modern necessity of keeping people informed." After rethinking the company's mission, Nick told the Conference Board, Time's management had revised its mission. Now its goal was to be a better buy on the stock market.

In other words, Time Inc. was more concerned with what a handful of Wall Street analysts and corporate raider specialists thought about what Time was doing than what Time's millions of readers and viewers thought about what it was doing.

I wonder what old Henry Luce would have thought about that.

Chapter 29

Home Video vs. Pay-Per-View

Too often the people in the entertainment industry have become obsessed with the way entertainment is delivered and not with the entertainment itself.

All of us love to be entertained, but we don't really care a great deal how that entertainment is delivered to us. In television, for example, the average American doesn't care how the picture got to the TV screen—broadcast over the airwaves or direct from satellite, cable, or videocassette. All that matters is the software—the program—not the hardware.

Yet the purveyors of television entertainment, just like the purveyors of plays, concerts, movies, variety shows, and so on, artificially divided the entertainment industry on the basis of hardware and delivery systems. We have the movie theater business, the broadcast TV business, the circus business, the stage business, the cable business, and the home video business—all fighting one another and all affected by the advancing technology of their little niche in the entertainment business.

It took years, but the cable people began winning the fight with the broadcast TV people when, due to Jerry Levin's vision, cable went up on the bird. Broadcast TV people were not paying attention to the technology and got caught by surprise, just as the movie studios had with talkies, radio, and television and just as the cable people had with the explosion of the VCR and the home video market.

What's at stake in all these frenetic activities is the $12 billion that Americans spend on home entertainment every year. It's a lot of money, and everybody is scrambling for a piece of the pie. According to the *Home Video Survey* made by the Fairchild Group, videocassette rentals grossed $2.35 billion in 1985 and almost double that, over $4 billion, in 1986. As of October 1985, *Home Video Publisher* reported there were 21 million homes with VCR units, up 6 million from the year before. By October 1986, that figure had grown to more than 26 million.

From the Hollywood movie studios' viewpoint, the home video boom was wonderful for two reasons. First, it broke the hold that HBO had on Hollywood. Many thought that HBO would own everything in Hollywood by 1990. Home video changed that by providing an additional major source of income for the studios.

Second, home video now means big money to the studios. Steven Rosenberg, an analyst with the Paul Kagan newsletter firm, predicted that videocassette sales would become the number-one source of income for the Hollywood studios by the start of 1987, and he was right.

The president of Times-Mirror Cable Company reported that the gross income figures for various kinds of television/movie-related business in 1987 would be: basic pay cable, $5.5 billion; premium pay cable, $4.3 billion; domestic movie theater box offices, $4.0 billion; and home video, $6.0 billion.

There was no doubt that the rapid growth of the home video and VCR market affected cable TV. People could become their own programmers, could do their own programming of movies and watch at their convenience. A major issue for premium programming services such as HBO and Showtime/The Movie Channel was what to do about this new contender for the consumer's entertainment dollars.

The choices were simple for the pay program services: They could come up with an alternative to home video, or they could jump on the home video bandwagon. The premium program services did both, and the cable operators, not being in the programming business, decided to come up with an alternative to home video, pay-per-view (PPV).

As we have seen, HBO responded to home video by getting into the home video business itself. Showtime responded in several ways. One of these was to get into the pay-per-view business just after Thanksgiving 1985. There were 36 million cable TV subscribers in America at the time of Showtime's announcement and 5 million of them had the equipment to subscribe to PPV.

Showtime had made a deal with Columbia, Paramount, 20th Century–Fox, and Warner Brothers to provide films for PPV that would be immediately available to 850,000 cable subscribers. The price for PPV would be set by the local cable company, but Showtime was recommending a charge ranging from $3.95 to $4.95 per film. Critics noted that this was more than double what most home video stores charged to rent a videocassette for one night.

In March of 1986, Showtime also joined HBO in launching a campaign about pay services's being VCR-friendly. In fact, VCR-friendly is the new 1986–87 buzzword of the pay cable world.

To the irritation of the movie studio people, who think VCR owners who tape movies should be shot, hanged, electrocuted, boiled in oil, HBO and Showtime/The Movie Channel are encouraging VCR owners to tape movies off their pay services.

For example, starting June of 1986, The Movie Channel began "VCR Theater" at 3:00 A.M. The idea behind the feature is to show a movie every morning that VCR owners can tape while they are sleeping. Peter Chernin, Showtime/The Movie Channel's executive vice president for programming, says that everybody knows subscribers are taping premium channel movies, and there is no reason to be coy about it. Instead he wants to give them another reason to keep subscribing to his pay service.

HBO hasn't gone quite that far, but in 1986 it did start an advertising campaign to cable subscribers telling them that HBO and VCR are a perfect combination for maximum home entertainment.

Jack Valenti, president of the Motion Picture Association of America, is livid at the prospect of people taping movies off pay channels. He says that no one should be allowed to tape copyrighted movies without permission and without paying a fee. Valenti had taken this position before, when he complained about people taping movies off of commercial TV networks, but the Supreme Court ruled against him in 1984. However, the Supreme Court didn't take a position on taping off pay cable, so Valenti is fighting the battle all over again.

Michael Fuchs's position is that there is simply no way to keep people from taping the movies HBO gets from the studios and shows on its service. Not to put too fine a point on it, Fuchs says, "HBO spends more money in the motion picture industry than any company in the world. Quite honestly, I don't need Jack Valenti telling me how to promote my business."

Chapter 30

The VCR Monster

In the minds of many cable operators, pay-per-view is the solution to the home video boom. They may or may not be right.

PPV allows the cable operator to charge subscribers for each event seen. As noted earlier, pay-per-view has been the golden dream of many cable people for a long time. It can, they believe, mean hundreds of millions of dollars in profit. PPV grossed $40 million in 1984, but a study by Arthur D. Little Company predicts it will reach $1.1 billion in revenue by 1990.

Mara Miesnieks, of Smith Barney, Harris, Upham & Company, said she thought that consumers were ready for PPV. Home video, she said, is pay-per-view on videocassette, only with PPV the consumer doesn't have to make two trips to the store.

Unfortunately, this is a case where the technology lagged behind the dream; the industry had experienced several false starts in PPV because the engineers hadn't perfected the necessary equipment. Beyond that, PPV meant that the cable operator had to invest in expensive new equipment. As a result the number of households that could subscribe to PPV was limited. But it was growing.

Besides movies, one of the strengths of PPV for cable operators is sports programming. Wrestling had been the mainstay of television back in the early days in the late 1940s and early 1950s, and history seems to be repeating itself. Wrestling has become a major feature of some PPV programming. In March of 1985 and April of 1986, such stars as Hulk Hogan and Andre the Giant participated in PPV's *Wrestlemania I* and *Wrestlemania II*. Combining PPV with VCR, wrestling also sold twenty thousand videocassettes of *Wrestlemania I*.

While some people groan and roll their eyes to the ceiling whenever TV wrestling is mentioned, the success of *Wrestlemania II* makes wrestling promoters smile. The April 7, 1986, *Wrestlemania II* was the third most successful PPV program in history, right behind the Cooney-Holmes and Leonard-Hearns title boxing matches. Heavily promoted, *Wrestlemania II* was seen by 360,000 homes on pay-per-view at an average price of $13.50 each. That's a total of $4,860,000.

Other sports promoters can do the math as easily as the rest of us. Regional PPV networks are already functioning, among them Home Team Sports (HTS) in the greater Washington, D.C., area, and Home Sports Entertainment (HSE), based in Dallas. By early 1986, HSE was showing some Dallas Mavericks and the Houston Rockets pro basketball games, plus games of both the Texas Rangers and the Houston Astros and about forty college athletic events. HTS has been offering some Washington Capitals hockey and Baltimore Orioles baseball games on a limited basis for $3 apiece. San Diego Cable Sports Network has been offering Padres baseball at $5.50 a game, and in Los Angeles "Dodgervision" has been offering PPV subscribers the games at $6 each or $85 for a twenty-game package. In its second season, 1986, Dodgervision attracted eighty-five hundred customers per game.

Other professional teams, such as the St. Louis Cardinals and the San Francisco Giants, are moving into the PPV market for 1987 and beyond on an experimental basis. Many experts regard sports as the best kind of PPV programming—or believe that it will be when PPV solves its technical problems. For example, MCA, which owns Universal Studios, tried to buy the Mets baseball team from Nelson Doubleday, and in September 1986 Bill Daniels, the biggest cable TV broker in the country, bought 4.99 percent of the Los Angeles Lakers basketball team.

Choice Channel's Rick Kulis optimistically predicts that by 1990 the pull of PPV money is going to draw many professional team owners into the PPV circuit. It combines the best aspects of TV for the owner, he says, while excluding the worst aspects of TV. PPV is an extension of the stadium in that it gives the owner more "seats" to sell without giving away the program on a flat rate.

In the February 1986 issue of *Marketing & Media Decisions* magazine Rich Zahradnik summarized the conflict between PPV and VCR. The VCR rental is less convenient, but movies are available on home video VCR cassette months before they come to cable TV and consumers have a wide range of movie titles to choose from—not just what the pay cable programmer picks.

The years 1985 and 1986 were poor for pay cable, with growth being much less than expected. The exploding home video phenomenon is given most of the blame, but is that accurate or fair? Jeffrey Reiss, president of the PPV Request Television which began in November 1985, thinks so. He points out that it's no coincidence that HBO and Showtime's worst year (1985) should also be home video's best year.

Reiss apparently doesn't put much stock in the effects of the sharply reduced construction of new cable systems and the way cable operators have alienated consumers with poor service and pay programmers have alienated consumers with repetition of movies.

There were four attempts at national PPV networks in late 1985 and 1986:

• Request Television, which started November 28, 1985, and has two hundred thousand subscribers. It does not act as the middleman between the movie studios and the cable system operator. Instead, it simply leases time on the satellite directly to the studios and lets them put on whatever movie they want.

• Viewer's Choice, a subsidiary of Showtime/The Movie Channel that began November 27, 1985, and acts as the traditional middleman between the studio and the cable system operator. That is, it rents the movie from the studio and rents the movie to the cable operator. As of February 1986, it had five hundred thousand subscribers.

• People's Choice, which began on January 3, 1986, with a showing of *Beverly Hills Cop*. It went under five months later, on June 1, for a very simple reason: not enough customers. People's Choice president, Lee Eden, figured that the PPV network would have had at least 300,000 subscribers by June 1, but it had only 125,000.

• Playboy's Private Ticket, which started in July 1985 and gives cable operators a three-hour tape once a week that is a combination of the Playboy Channel and movies. The system operator plays the tape as often as he wishes each week. It has 150,000 subscribers.

There are three other PPV services being talked about—Telestar, EventTelevision, and Choice Channel—but they hadn't gone ahead as of the beginning of 1987. EventTelevision is jointly owned by six cable conglomerates and Caesar's World in Las Vegas. It expects to have eight hundred and fifty thousand subscribers, but it hasn't decided just when to get started. Sid Amira, the executive director, is bullish on PPV and predicts that people will pay more for a PPV movie—an average of $4.95—than a rental video because of the convenience of not having to pick up and return the cassette to the home video store.

John Hunt, vice president of media research for Oglivy & Mather, and Ira Tumpowsky, senior vice president for cable at Young and Rubicam, disagree. They believe that people enjoy going to the video store, mixing with people, and browsing through the selection of movie titles available. In addition, Tumpowsky notes that, with a videocassette, the consumer can stop and start the movie at will and even replay it several times.

Ironically, Frank Biondi, formerly president of HBO and now executive vice president of Coca-Cola's Entertainment Business Sector, is a major advocate of PPV. One of the main points he makes to movie studios is that PPV is better for their business than the home video market. As he explains it, once a studio sells a VCR cassette to the home video market, that's it. It never gets a piece of the rental income from the cassette. However, with PPV, the studio can get a piece of every showing of the movie. That's why he thinks the studios ought to release movies to PPV before they release them to the home video market. Most experts agree that the one thing critical to PPV's success is this early availability of movies to compete with home video.

Other problems exist, of course. For example, PPV means another major hardware investment up front by the cable operator. It is estimated that this investment runs $150 to $200 per subscriber. For example, in San Antonio, Rogers Cable TV has spent $7 million to give its forty-five thousand subscribers a PPV converter. That's a cost of $156 per subscriber. However, Viacom Cable built a brand-new cable system in Nashville to include PPV. It cost $47 million for sixty thousand subscribers, or $783 per subscriber. If a cable operator collects an average of $5 per month and gets to keep only half of that, it will take a long time to recoup the investment.

Another major problem is the wide variety of equipment available for PPV. The ideal PPV would be a two-way cable that would allow the subscriber to communicate a PPV order to the cable operator over the same wire that would deliver the PPV program. However, this technology turned out to be too complicated and too expensive.

The alternative most widely used now is for the subscriber to call in his order on the telephone and then have it transmitted over the regular one-way cable to a special converter box in the subscriber's home that scrambles the picture unless it is instructed by the cable operator to unscramble it.

According to Paul Kagan, there were only 175,000 homes with these converters in 1981, but that number has grown to 9 million today and is projected to be 37 million in 1994. The January 1987 A.C. Nielsen survey said there were 87.4 million American homes with television sets, and that 39 million of these were connected to cable.

The problem with the telephone order system is that PPV subscribers normally decide on impulse and at the last minute to order a PPV movie. This means jammed telephone lines into the cable operator's office, endless busy signals, and frustrated customers. It is a proven fact that consumers quickly reject complicated or frustrating methods of fulfilling their desires, so unless PPV finds a quick and easy system of ordering and delivering PPV, it could be a bust.

Cox Cable San Diego has a system that may represent a step in the right direction, but it's not widely used. The Cox system allows people with Touch-Tone phones to use Touch-Tone signals to talk to the Cox Cable computer and order the desired movie. It takes about thirty seconds, but it still relies on having enough telephone lines available to handle the incoming calls. For example, on opening day of the San Diego Padres' season in 1985, 404,000 phone orders flooded in and jammed the lines.

Other methods of speed-ordering are being tried, such as the Rogers Cable systems, which let people use their remote control signalers to talk to the computer; and Telescripts Industries, which has developed an electronic chip that you use like a credit card in an automatic-teller machine.

Even so, pay cable programmers are worried that PPV will hurt their business. HBO has not decided what its position will be on PPV, but Michael Fuchs has already said he is worried PPV will "cannibalize" pay program services. Some Wall Street analysts disagree, saying that PPV and pay services can coexist comfortably. Of course, some pay services—Showtime/The Movie Channel, for instance—are getting into PPV as fast as they can.

Fuchs does think that the time may come when HBO will get into PPV. In February 1986, he told the Washington Metropolitan Cable Club that he saw a time for PPV but that it was several years away. His reasoning was that the technology was lagging behind the marketing people and the industry had already given the consumers as much as they could handle at present. Some cynics say that HBO is holding back from PPV because of its partnership with Paramount and Universal Studios in the USA Network. The arrangement freezes HBO out of PPV unless the partners are included.

There are major cable operators who say they don't want PPV at all. John Malone, president of the largest conglomerate of cable systems in the country, Tele-Communications, Inc., says that TCI doesn't have any PPV operations and doesn't want them. In fact, where TCI had bought a cable system that came with PPV, it killed the PPV.

Malone says the reason is that it is too expensive to operate for the revenue it brings in. The most fundamental marketing lesson of cable television is that the cable operators will decide what to put on their cable systems, not the programmers. No matter what grandiose ideas some programmer has, he still has to convince the cable operator to let him on one of the operator's channels. Of course, the cable operator is interested primarily in what is good for *business*, not what is good for the programmer.

It is curious that, in spite of what John Malone says about PPV's not being worthwhile for TCI, he has a strategy for putting PPV on his cable systems anyhow. The strategy is designed to ease the pain of the additional capital investment in hardware for the cable operator while appearing to bring PPV charges more in line with home video rental rates. It would also eliminate the middleman, because TCI would deal directly with the movie studio.

The Malone strategy is to charge subscribers $6 a month for PPV plus $2.50 for one movie a weekend shown continuously. Under the Malone plan, TCI would keep the $6 plus 25 cents per movie. The remaining $2.25 would go directly to the movie studio. (That's what the movie studio gets now when one of the PPV networks acts as broker between the studio and the cable operator.)

As *Forbes* magazine reporter Jeffrey A. Trachtenberg noted in the August 26, 1985, issue, savvy cable operators must understand that their best move in 1987–88 will be to get more subscribers to their basic cable services. At the beginning of 1987, approximately 60 percent of the homes with TV cables passing them were not cable subscribers. Without investing in more hardware, they can increase their income. What's more, the Congress has socked it to the cable subscriber by deregulating cable as of January 1987, and most cable subscribers can expect a 50 percent jump in their monthly fees. So, what is to be gained by experimenting with expensive PPV?

Also, as Trachtenberg observes, HBO and Showtime/The Movie Channel are pretty good bargains. The consumer gets thirty-five to forty movies a month for $10 to $15. For the same average price of PPV, one can only see two or three movies. Seth Abraham, HBO's senior vice president for programming, underscores that by saying that nobody can get 142 movie titles from the video store or from PPV. "That's what works for us," he says, "sheer volume of titles."

Another problem is the same one that has plagued the cable programmer for years: Where to get good programs? Ten percent of the movies released every year account for 45 to 50 percent of the box office revenue, so consumers might pay $4.95 for a movie blockbuster that has been hyped by the studios for theatrical release, but they won't pay it to see *Howard the Duck*.

In the conflict between cable and home video, one important fact is ignored by many. Cable still has a lot of potential customers if it gets its marketing act straightened out. Some 40 percent of the homes passed by cable still do not subscribe, and one in four television homes cannot get cable because it hasn't come to their neighborhood yet. Between those two situations, there are about 47 million more, homes that are potential cable customers.

According to Paul Lindstrom, research vice president of the A.C. Nielsen Company's Home View Index, the market for the VCR is much more limited. He says there will come a point when VCR sales "hit the wall" and stop dead. The VCR user tends to be a family making $20,000 or more, and a high percent of families in that category already have bought VCRs. Lindstrom reports further that 35 percent of all American homes had a VCR as of February 1986. Of the families making $20,000 or more, 87 percent or more already had VCRs by the end of 1986. Therefore, he expected the VCR boom to taper off by the end of 1987, and the figures support him as VCR sales have leveled out.

Another warning signal is the total number of VCRs sold. While sales had been good in previous years and were still good in 1985 and 1986, they peaked in 1984. That was the year that pay TV hit a plateau in growth.

In 1983, 4,091,321 VCRs were sold. That rose to 20,991,732 in 1984, which was the record of VCR sales to date. In 1985, VCR sales were 20,749,527, but projections for 1987 are only about 15,200,000.

Kagan Associates put on a seminar in Los Angeles in mid-1986 to discuss the relative merits of home video versus pay-per-view. One of the major points to come out of the seminar was the Hollywood studios' long-term faith in the future of pay-per-view. Even though home video is the hot performer at the moment while pay-per-view sputters along trying to get up to speed, Hollywood studios keep looking to PPV as source of massive profits in the years ahead.

Something else that became clear in the Kagan seminar was that, on the issue of VCR vs. PPV, there seem to be as many opinions as experts. Some say it will be a $5 billion business by 1995; others say it will be a dead duck by that date. In between, there are scores of varying views.

In the face of all the disagreement, here is one hard fact: *In 1987, home video was the largest single source of money for the Hollywood studios.* It paid Hollywood $2.3 billion, as contrasted with the paltry $70 million that PPV brought in.

The big stumbling blocks to PPV are the reluctance of cable operators to spend the money to install the necessary equipment; the problem the PPV equipment manufacturers are having developing and producing dependable equipment; and the resistance of many consumers to pay for each movie seen.

For the studios, the economics of home video are not as attractive as the economics of PPV. A big movie in videocassette will sell a maximum of 250,000 copies at $50 wholesale and $79.95 retail. The studio gets part of that $50 after the production cost and distribution costs are deducted—that might be around $10 for the studio. From home video rentals, the video store will average ninety rentals per film and make an estimated $15 or $20 net, but the studio will often get nothing. With PPV, a big movie might be seen in two million homes at $5 a home with the studio and the cable system splitting the take evenly, or $5 million.

Premium programmers complain that the studios release films six months earlier to home video than to cable, and this, of course, hurts cable. However, Dave McDonald of New Jersey's *New York Times* Cable says that earlier release doesn't help HBO because the people who disconnect usually do so after having HBO over a year. They have seen most of what HBO has to offer and watch it less and less until they finally disconnect.

McDonald spends more than $100 per subscriber to make it possible for 32,000 of his 134,000 subscribers to receive PPV. In April, 28 percent of his basic-only subscribers took at least one PPV event. By 1990, he expects to sell 150,000 PPV tickets a month.

In the end, technology continues to fragment and confuse the entertainment industry, because it complicates the way entertainment is delivered to the consumer. Of course, the consumer doesn't care about technology; he just wants convenient entertainment at an affordable price. HBO and Showtime/The Movie Channel understand that, and each of them has made a counterattack on the home video market.

HBO launched a campaign urging people to subscribe to HBO and tape movies off the channel to build a home video library for themselves. Typical of the commercials HBO is running is the one that reads,

> Record HBO and cable programs with your VCR and you'll see what you want, when you want. Or start your own video library! You won't have to face any more travel headaches or out-of-stock disappointments at the video store.

The Movie Channel has switched from monthly to weekly rotation of its movies, added a host, and created the "VCR Theater." It shows some of its important films at 3 A.M. so VCR owners can tape for later viewing. The "VCR Theater" host will tell viewers how best to use their VCRs.

Naturally, James Bouras, vice president of the Motion Picture Association of America, was unhappy about this new scheme. Jim English, vice president of pay TV at MGM, said The Movie Channel's plan was shocking and "shows no respect for Hollywood." He expected MGM would jump its prices by 50 percent as a result.

Michael Fuchs told a Washington, D.C., audience of cable people back in September of 1986 that he saw PPV as being far down the road because of the overcomplicated technology. He believes many cable systems are moving away from the expensive equipment needed for PPV in favor of simpler, more reliable equipment. Fuchs thinks that home video has hit the wall. He means that, in spite of the supposed growth in the number of VCR machines sold, the home video retailer is still ordering about the same number of videocassettes of each film.

The retailer doesn't want to tie up a lot of his money in a big inventory of the blockbuster movie cassettes. Most retailers assume people will come into the video store to rent the latest hit movie, but if that isn't available because of a limited inventory of hits, the customer will rent another, less desirable movie. (Of course, in doing this, the retailer is liable to make the same mistake others have done in the past. For example, pay cable television attracted subscribers by promising hit movies, but it didn't deliver them every night, and people began disconnecting.)

Fuchs is fearful that PPV will erode HBO, but admits that in Honolulu's system (owned by a sister Time Inc. subsidiary), PPV hasn't hurt HBO.

Meanwhile, Tele-Communications, Inc. (TCI), the biggest cable system operator in the country, started in early 1986 to test PPV in a few of its systems while it negotiated with Hollywood to get movies much earlier than they were released to cable systems normally, so that subscribers could watch or tape them. The concept is called "Movies on Demand" and uses simpler, one-way equipment. The idea is to encourage cable subscribers to build their own video tape libraries and woo them away from the home video stores.

John Malone, president and CEO of TCI, said that 65 percent of home video store sales are of old movies, because there are not enough new releases available. The "Movies On Demand" concept would tap into this market.

While insiders argue about technology, HBO and other programmers remain flexible enough to jump into whatever technology will work for selling their programs to the consumer.

Chapter 31

Some Little Piggies to Market Wouldn't Go

Marketing has always been a problem in the cable industry, largely because the cable operator has shown over and over again that he would rather skim the cream, ignore servicing customers, and keep as much of the revenue as possible instead of investing it in promotion and marketing. Naturally, this has a bearing on HBO's success, because HBO depends on cable operators to sell its service.

For both the cable industry and HBO, 1985 was a poor year, because of problems in the cable industry's construction plans and because of the explosion of the home video/VCR market. A 1983 HBO report projected it would have 18.8 million subscribers by the end of 1985. That was too optimistic, but HBO came close to achieving that goal by the end of 1986.

In 1985, HBO gained a hundred thousand new subscribers; and Cinemax, 400,000, for a combined total of five hundred thousand. Seth Abraham, senior VP for programming and sports at HBO, said it was because HBO programming is "videoproof"— that is, there are programs, such as sports, specials, and movies— that you cannot get on home video. However, when home video became a threat to HBO, it formed Thorn EMI–HBO Home Video and got into the home video business itself. One of its big home video hits was *Amadeus,* which sold over 150,000 and made over $12 million. (It is an indication of how HBO is involved in every form of entertainment that it decided that the asking price of $8 million for the pay cable rights to *Amadeus* was too much and it passed on it.)

In 1984 and 1985, a lot of people thought that it looked as if the wunderkind of Time Inc. had fallen on its face suddenly and without explanation. That, as we have seen earlier, was partly due to myopia in the HBO executive suite and partly due to sharp changes in the cable industry. It was really not all that bad, it was just that HBO wasn't as good as it had been before. It was no longer living up to its press clippings.

HBO was hot for a long time and was seen as a monster about to devour Hollywood. HBO is still very big; it rescued the cable industry in the mid-1970s and is still an important element of the cable industry. However, HBO is no longer just in the cable business, it is in the entertainment business competing in cable, network TV, home video—everything. It owns a piece of USA Network, the Black Entertainment Network, Loews movie theaters, several movie production companies, and on and on.

HBO has gone through two tough years, partially due to competition from VCR and home video and partially due to the reluctance of cable operators to continue building new systems and to cable operators' sluggish marketing.

And as CEO Michael Fuchs told *Broadcasting* magazine in September 1986, people at HBO began believing what everybody else was saying about HBO's being on the fast track when, all of a sudden, "the caboose began slamming into the front of the train."

Even so, HBO is the biggest profit center Time Inc. has; it makes more profit than any individual magazine in the Magazine Group. It is the most profitable company in the cable industry.

People talked about pay TV for two decades without success, but HBO made it happen. Then HBO came out of nowhere with nothing and became the biggest monthly brand product in the world.

As Michael Fuchs sees it, one of HBO's main problems has always been that its primary distribution system is through cable operators who are notoriously poor marketers, spending less than 1 percent of sales revenue promoting their businesses. There is no other element of the entertainment business that spends so little on promotion.

What is needed is innovative marketing to overcome objections of the skeptical never-subscriber consumer.

Walter Giera, General Manager of Cablevision in Uniontown, Pennsylvania, says that cable has to wake up to the reality that cable TV isn't the only entertainment option to commercial TV in town. There are VCRs, movies, and commercial TV, and cable operators have to merchandise to survive.

Giera thinks cable has a bad reputation to overcome. He tries to accomplish that by teaming up with companies that have good reputations. For example, he has tried grocery store displays and giving away Coca-Cola for subscribing to basic cable or to HBO. He avoids newspaper ads and relies instead on direct mail to reach the audience that doesn't read newspapers but does shop.

Ernie Quinones, Group W market man in El Paso, says that much can be done even in the face of the economic crisis of Texas. He tries all sorts of merchandising techniques, giveaways, and promotion. He sells cable as the cheapest entertainment buy.

One very successful approach for getting to the hard-to-find, hard-to-reach consumer is an HBO-sponsored program that offers its affiliates extended free-trial programs of basic and HBO and Cinemax. HBO vice president Matt Blank says that the traditional two or three days of free previews has proven effective in the past. The free ten-day trial uses the same approach to target cable resistors. During a test of this approach, HBO says there was a 59 percent jump in HBO and Cinemax subscriptions.

Of these taking the ten-day free trial, HBO reports that 60 to 80 percent kept the basic service; 30 to 60 percent stayed with HBO; and 25 percent stayed with Cinemax. It costs HBO about $30 per subscriber for the free trial, but HBO sees that as a cost-effective expense in a maturing market.

The most impressive promotion was run as Prime Ticket for Comcast system in Tustin, California. It included a big sports package of Los Angeles Lakers games, boxing, tennis, and tape-delayed games of local colleges. There were also prizes, jackets, T-shirts, and other giveaways. Basic subscriptions soared by 22,500 in nine weeks.

To help cable operators become more effective merchandisers, HBO has started an in-house telemarketing training program, called ProToCall. It is for cable operators who want to run their own telephone marketing program. Barbara Jaffe, HBO VP of sales promotion, said, "Telemarketing is one of the fastest-growing marketing tactics in the cable industry today." It gives operators a way to get new subscribers or to get old subscribers to buy new services. The idea is to teach cable operators' salespeople how to sell on the telephone.

In spite of these and other efforts, the cable industry really hasn't sold the public on what a good buy cable is. Since the punch of first-run movies has been taken away by home video, HBO may have to compete on the basis of value and volume. For example, HBO at $10 a month delivers a hundred movies and fifty special programs each month.

HBO subscription levels are coming back up a bit. Michael Fuchs says that 1986 was a better year than 1985, with subscription totals at about 14.6 million on HBO and 3.7 on Cinemax. In fact, HBO/Cinemax was the only twenty-four-hour pay service to show growth in 1985 and 1986.

On the volume side, HBO is not convinced that any movie is worth spending huge amounts of money to produce, so it is going for more movies at lower individual cost. HBO is producing more original programming, movies, specials.

But Fuchs worries about greedy cable operators who don't give subscribers value for their money. He says that the HBO experience is that it is very hard to get people who have churned out to come back. "Once you scorch the consumer, it's very hard to sweet-talk them back in," says Fuchs.

Fuchs is disappointed in the cable industry; he thinks HBO is being sold and marketed all wrong. There should be better and more promotion. He is frustrated that the cable business botched up their franchises, overpromised, cut back, stopped building, took the easy money, and ran. He would like every home in America wired.

Chapter 32

HBO Programming 1986–88

Programming has been and always will be the heart of the pay TV business, something that HBO has known all along. A variety of hardware delivers that programming, but it's the programming that's key.

As HBO grew wealthy and hungry for more programming, it found that it couldn't find enough to fill its needs from outside sources. That's what initially drew HBO into doing a lot of its own programming, first by financing movies and special events, then by producing movies, concerts, sporting events, and other specials either alone or in cooperation with partners.

Today, HBO is the country's largest financier of movies and a major producer of special events and sports happenings. It is involved with home video and it holds theatrical rights to many movies and programs, but cable is still the core of HBO's business. In this chapter we'll see what's in store for HBO viewers in 1988 and beyond.

First, there is an anthology series of sixty- and ninety-minute programs not unlike the old "Playhouse 90" or public broadcasting's "American Playhouse." Called "HBO Showcase '86" and "HBO Showcase '87," the series touches on a number of different subjects. For example, *Nosenko*, starring Tommy Lee Jones, is the story of the KGB defector Yuri Nosenko. Another production, *Half a Lifetime* with Keith Carradine and Gary Busey, shows what happens when some poker-playing buddies face a crisis. There are Liza Minnelli, Olivia Newton-John, and Barbra Streisand concerts and comedy programs, such as the strange Max Headroom and Cinemax Comedy Experiment. Less strange comedy is also on HBO as part of its "On Location" series. Comedians in concert include Buddy Hackett, Robin Williams, George Carlin, Billy Crystal, Howie Mandel, and Richard Belzer.

While cable hasn't delivered the bonanza of innovative and high-quality programming that was once predicted, there have been some hits, particularly in comedy presentations. Lee Margulies, critic for the *Los Angeles Times,* was particularly impressed with what he dubbed the best comedy special of 1986, "Robin Williams—An Evening at the Met." He said,

> You need a seat-belt to stay with Williams' wild, stream-of-consciousness ride through life in the '80s—from President Reagan to TV evangelists, from international politics to sex, from pollution to pop culture, from drugs to fatherhood.

HBO will probably also have more sneak-preview releases of major blockbuster movies. For example, back on Christmas Eve of 1985, it promoted an "HBO Christmas Present" for its subscribers. It turned out to be *Ghostbusters,* shown only once, at 11:30. This was only two months after *Ghostbusters* had been released to the home video/VCR market and several months earlier than it would normally have been shown on cable TV. In order to make the deal with Columbia Pictures for this special screening, HBO was not allowed to announce it in advance. (Columbia didn't want everybody videotaping the movie. It had already become the biggest-selling videotape in the $79.95 price bracket, with four hundred thousand sold, and Columbia hoped it would sell a lot more.)

Some people think that, while HBO made a lot of people at home happy with this Christmas gift, it also probably made a lot of its subscribers angry because they didn't know about it in advance and may have missed it.

In another movie deal, HBO put $10 million—about half the cost of the film—up front into the comedy *Three Amigos,* with Steve Martin, Chevy Chase, and Martin Short. In exchange, HBO gets the exclusive cable TV and home video rights to the movie forever and a piece of the profit from its theatrical showing anywhere in the world. It had a similar deal with *Volunteers, Sweet Dreams,* and *Head Office.*

At the beginning of 1986, HBO signed a three-part deal with

20th Century–Fox. The first part of the deal gives HBO the nonexclusive pay rights to Fox films produced between 1985 and 1988. These include *Commando, Cocoon,* and *The Jewel of the Nile,* plus a package of fifty-five older films, including five Charlie Chaplin movies. The second part of the deal is for cooperative financing of ten new films to be made by HBO Premiere Films. Fox gets the theater exhibition and home video rights, and HBO gets the pay TV rights.

The advantage to HBO of this is that it will produce more films for its cable programming services for fewer dollars because the cost is shared with Fox. This is particularly important to HBO since the pressure is on from its parent, Time Inc., to cut costs and get more productivity from its investments.

The third part of the deal is for Fox to distribute HBO's theatrical films produced by such HBO production companies as Silver Screen Partners and Tri-Star Pictures.

Two deals made in the summer of 1986—one between the movie studios and Showtime and between the studios and HBO—underscored the different approaches used by the two biggest cable program services.

HBO made a nonexclusive deal with Warner Brothers for Warner films to be produced in the next five years. Some of the titles include: *Under the Cherry Moon,* with Prince; two Steven Spielberg movies, *The Goonies,* for the acne set, and *The Color Purple,* an adult drama; and two comedies, *Police Academy 3* and *Spies Like Us.*

This five-year package, which will cost HBO half a billion dollars, supplements an earlier deal that included *Pale Rider,* starring the new mayor of Carmel, California, Clint Eastwood, and Spielberg's *Gremlins.*

Then, in August 1986, HBO made another big deal with MGM-UA for $300 million that gives it the nonexclusive cable rights to seventy-two movies, including *Rocky IV, Youngblood, To Live and Die in L.A., Running Scared,* and *Poltergeist II.* From one of its own studios, Premiere Films, HBO shows eight or ten films a year, including *As Summers Die* (starring Bette Davis in a drama set in the South during the 1950s) and *Apology,* a Lesley Ann Warren film about a New York artist who is stalked by a killer. There are also specials, such as the three-hour session of "Ray Bradbury Theater" starring Jeff Goldblum, Peter O'Toole, and Drew Barrymore.

162   *Inside HBO*

Because HBO gets these on a nonexclusive basis, Warner is free to offer them also to other cable operators, including Showtime, and regional cable systems such as The Z Channel in Los Angeles and the Prism Channel in Philadelphia.

Soon after HBO made the Fox deal, Showtime/The Movie Channel made a $400 million deal with The Cannon Group, Atlantic Releasing, De Laurentiis Entertainment Group, and Walt Disney Studios for 140 films, including *Runaway Train, The Delta Force, Death Wish III* and *IV, Echo Park* with Tom Hulce, *The Men's Club* starring Roy Scheider and Treat Williams, *Extremities* with Farrah Fawcett, *Down and Out in Beverly Hills,* Paramount's *Beverly Hills Cop,* and the star-laden *Crimes of the Heart.*

In June and July of 1987, HBO made two giant movie deals, one with Coca-Cola Telecommunications and the other with Paramount Studios, which has been working toward the end of its five-year exclusive deal with Showtime.

The $70 million Coca-Cola deal (Coca-Cola owns Columbia Pictures) calls for fifteen made-for-cable TV films. These will be shown on HBO cable TV, and HBO will also have the home video rights to ten of the films.

The Paramount deal is much bigger, involving eighty five films for half a billion dollars on a five-year exclusive deal starting in 1988. The deal, similar to the one with which Showtime had stunned HBO five years earlier, also stunned Showtime. Analysts believed that it was a deal that would hurt Showtime, but not fatally.

It is estimated that Showtime/TMC has about 8 million subscribers as of 1987, with HBO/Cinemax having a combined total of approximately 19 million. While home video has been a strong competitor to cable, cable subscriptions were rising again in 1987 due to extensive promotion campaigns.

The New York Times of May 24, 1987, commenting on the HBO/Paramount deal, reported that one industry analyst thought the deal would hurt Showtime seriously and leave HBO with a virtual monopoly in the pay television business just as it had six years before.

As I noted earlier, HBO had relaxed its position on exclusivity, in spite of the Paramount deal, but Showtime was insisting on it. In theory, then, a subscriber who wanted only one premium cable channel might be better off going with Showtime because there are movies that he can see only on Showtime; theoretically he could see the nonexclusive movies that HBO has because Showtime could get those same movies if it wanted to do so.

But there's a catch in the Showtime strategy. Showtime says it will not take a movie unless it has it exclusively, which means that it refuses to show movies on which HBO has made a deal even if the deal is nonexclusive with the studio. In essence, Showtime is making HBO's nonexclusive deals with the studios exclusive deals.

That's okay with HBO's Steve Scheffer, executive vice president for film programming and home video, because nonexclusive deals cost less than exclusive ones. Besides, Scheffer thinks that exclusivity can lead to locking up a lot of second-rate films and not having access to some of the blockbusters that attract subscribers.

In October 1986, HBO went to the Securities and Exchange Commission (SEC) to get permission to do another massive film-financing deal. Silver Screen Partners, an HBO film-financing operation, would raise $300 million from limited partners to finance eighteen new Disney films just as soon as the SEC approved. The planned films include Charles Dickens's *Oliver Twist* as an animated musical. In 1985, Silver Screen had raised $200 million to finance such Disney films as *The Great Mouse Detective*, *One Magic Christmas,* and *Tough Guys.*

The Silver Screen Partners idea, created by former HBO president Frank Biondi, allows many small investors to put $5,000 or more into film production. While film production has always been a risky business, Silver Screen Partners has had a very good profit track record.

Attracting subscribers is the name of the game in pay cable, and it is a game at which HBO has excelled so far. In spite of slowing growth during 1984–1986, HBO still has half again as many subscribers as all the other major premium programming services combined. At the end of 1987, it had about 19 million subscribers.

On the sports side, HBO continued its dominance of heavyweight boxing on cable by signing a $20 million agreement with boxing promoters Butch Lewis and Don King. This was for a series of seven title matches over 1986 and 1987, including fights between Pinklon Thomas and Trevor Berbick as well as between Michael Spinks and Larry Holmes.

While all of these programming deals are going on for HBO and Cinemax, HBO is quietly going ahead with a third programming service called Festival. The programming focus of this new service is adult programs without excessive sex or violence. For example, movies such as *Places in the Heart, The Natural,* and *The Karate Kid* would be shown on Festival and it would also run historical and nature documentaries. Festival would also show some R-rated movies, such as *The Killing Fields* and *Frances.*

HBO sees the audience for Festival at a relatively modest level, of about 1.5 million subscribers, mostly older and mostly turned off by pay television up to now. The first step was to test the market for Festival, and HBO has tried it on forty cable systems. In order to be judged a success for HBO, the Festival service will have to make money with relatively few subscribers and it will have to get subscribers without stealing them away from HBO or Cinemax.

All of this is only a sample of what is going on at the two major pay cable program networks. The point is that both Showtime/The Movie Channel and HBO/Cinemax are deeply into financing and producing a wide range of programs. In 1986 alone, HBO committed well over $1 billion dollars to movie production.

It is also why HBO is such a cash machine for Time Inc., generating 46 percent of that company's *net* profit even though the growth in cable TV has flattened out, and there is competition from other pay services, home video, and pay-per-view.

HBO has become the biggest entertainment conglomerate in the world, and it is getting bigger. Motion picture and television mergers are producing an enormous concentration of power in a handful of companies. David Londoner, vice president of the investment firm of Wertheim & Company, predicts that three major entertainment companies will control 80 percent of all broadcasting and entertainment programming within the next ten years. Sidney Sheinberg, the president of MCA/Universal studios, agrees.

Movie studios are buying theaters; programmers are buying or creating studios; broadcasters and cable operators are into studios. For instance, Ted Turner bought MGM and, in turn, sold off pieces of it and his own company in 1987 while Rupert Murdoch bought 20th Century–Fox. Little operators are being bought up or squeezed out. In the booming home video business, for example, the studios are feeding their movies to their own video divisions and freezing out independent distributors more and more.

Cable programmers such as HBO are into the home video

market, into movie theaters, into film production, into about everything they can lay their hands on.

One reason for the insatiable appetite that companies such as HBO and Time Inc. have for buying up other related companies is Ronald Reagan. The Reagan administration has been very pro big business and pro big profit, but the tendency is for government regulation to tend toward the middle over time. A move back to tighter regulation might be coming in the next few years. HBO's Michael Fuchs says that something he hears a lot these days is "Do it now. There is only one year of Reagan left."

The entertainment business has been through all of these possibilities before. In the days before the end of World War II, the movie companies owned the film making facilities, the distribution system, and the movie theaters. Then, along came television, which the movie companies tried to force the government to ban or regulate heavily. The government didn't, but the government did make the studios get out of the movie theater business.

Starting in the early 1950s, things settled down to a new balance of power between broadcasters and movie studios in the battle for the consumer's entertainment dollar. Then, in the early 1970s, cable television came along. First the federal government didn't think it should regulate cable. Then the broadcasters screamed about the competition, and the government changed its mind and almost regulated cable out of business.

Things settled down to another uneasy truce among cable, broadcasting, and the movies until another technical advance came with the VCR boom and the deregulation of government controls over broadcasters and movie studios. Now everybody is trying to get back into everything it used to be in plus a lot of other things, too.

Now, in the late 1980s, merger mania has hit the entertainment world, and things are going so fast, it is hard to keep up. A few months ago, during a very complicated buyout of MGM by Ted Turner, the joke on the studio lot was, "If you see my boss, call and tell me who he is."

Some experts—such as Bruce M. Owen of a Washington, D.C., consulting firm, and Herbert Allen, Jr., of Allen & Company (the firm that used to own Columbia Pictures)—say that a lot of entertainment industry deals are now being made from a sense of previously lost opportunity. Everybody wants to make a deal for fear he will regret not doing so in the future.

One of the things that concerns Sid Sheinberg of MCA is that little producers may get squeezed out, and program censorship or blackouts may occur when a few large companies again own the means of production, distribution, and exhibition of entertainment. How could any such company not favor its own material over that of an outsider?

Whatever happens, HBO and Time Inc. will be among the players. HBO, troubled as it has been, still generates a lot of cash profit for Time Inc. And, as baseball manager Casey Stengel might have said, "the bottom line is the bottom line."

Chapter 33

Issues and Maybe Answers

For the rest of the 1980s and the 1990s there are several issues that will affect the cable industry and the growing number of Americans who subscribe to cable TV.

In the short run, scrambling will be a major issue. In the medium run, deregulation will be a major issue. In the long run, the major issues will be international expansion of cable TV; getting more and cheaper programming; the continuation of the technology explosion, but at a faster rate, which may soon bring us commercially feasible Direct Broadcast for Satellite (DBS) and such other goodies as High Resolution Television, which allows giant screens with razor-sharp images and digital sound. In the short, medium, and long runs there will be the constant fight over who gets how much of the consumer's entertainment dollar.

As we noted earlier, one of the great issues irritating cable operators and program suppliers has been pirates—the people who buy their own receiving dishes and pull programs off the satellites without paying for them. No one in the cable industry has an accurate figure of how many people steal these signals. However, my friend Ed Dooley, who was Director of Public Information for the National Cable TV Association, estimated that the industry was losing hundreds of millions of dollars every year to pirates. State and local laws tend to be weak on this point, and law enforcement officers feel they have a lot more important things to do than to root out cable TV pirates. In fact, the industry was surprised a couple of years ago when three New York bar owners were indicted and convicted of cable TV piracy.

167

In any case, HBO decided some time back that the only way to scotch the pirates was to scramble their signal so one needed a descrambler to view it clearly.

In my days at HBO, meeting after meeting was held to plan how we would prepare the public for what we referred to around the Time-Life Building as "The Day the Skies Would Go Dark."

One of the reasons is that there are tens of thousands of people in America who have paid a lot of money to buy their own television receiving dishes (called a TVRO for TV Receive Only). While a lot of these people are blatant pirates, many are in areas not served by cable TV or good broadcast television. Isn't it immoral and unpatriotic to deny people their favorite programs? Any time you do something that affects a lot of people, there could be a serious backlash.

Well, on January 15, 1986, HBO finally got around to scrambling, and it did make a lot of people angry. Showtime said it was still testing and would scramble later, but most people think that Showtime was just letting HBO take the initial heat. If scrambling worked out for HBO, Showtime would jump in and do it, too.

It cost HBO $15 million to scramble its picture, but it tried to make a deal with TVRO dish owners offering to sell a descrambler for $395 each and give them HBO for $12.95 a month and $19.95 for both HBO and Cinemax.

At those prices, HBO didn't expect many sign-ups from TVRO dish owners, but surprisingly, the rate of descrambled customers got up to about seven hundred a day within a few weeks after HBO began scrambling. About 90 percent of these TVRO customers are subscribing to both HBO and Cinemax. In the meantime, not surprisingly, there has been a huge drop-off in the sale of TVRO dishes. (SPACE, the Satellite Association, expects to sue for restraint of trade and other things.)

At the 1986 annual convention of the National Cable TV Association in Dallas, attending congressmen told cable operators and program service people that Congress saw scrambling as a public issue. They warned that the dish industry would die because of scrambling unless cable and dish industries began cooperating and said that Congress might reluctantly get drawn into doing something about it if the cable industry didn't.

Meanwhile, the Department of Justice was stepping up its probe of scrambling to see if it constituted a deliberate attempt to destroy a competing industry. Many prominent organizations involved in scrambling such as HBO, ATC, NCTA, TCI, and Viacom were issued Civil Investigative Demands (CIDs) by the Justice Department to see if there were a restraint of trade or other violation of federal law.

Representative Tauzin (Democrat of Louisiana) warned that while Congress was reluctant to get involved in the scrambling issue, it would if some agreement wasn't worked out. What Tauzin, Sweeney (Republican from Texas), and Billey (Republican of Virginia) wanted was to have descrambling kits available to dish owners at prices no greater than what cable subscribers paid for cable TV.

The House Telecom Subcommittee's senior counsel, Thomas Rogers, said he thought scrambling involved "significant public policy concerns" that Congress would deal with unless the marketplace took care of it soon. In addition, there were the matters of service to remote dish owners beyond broadcast signal range and outside cable franchise area; consumer retail dish sales; privacy of business data delivered by satellite; and shifting of programs to the Ku-band on satellite that is incompatible with most TVRO dishes.

Some were not content to wait for Congress to act or the cable industry to cooperate. Lawsuits started getting filed in the spring of 1986. For example, several pay-per-view services, such as Personal Preference Video and its distributor, Space Age Video, filed suit in U.S. District Court, Fort Worth, against Paramount, Universal, and HBO. The suit claims restraint of trade because HBO and others have refused to provide adequate descramblers to dish owners.

Robert Huff, chairman of the dish dealer's United Satellite Association, said they would soon file a similar lawsuit. They wanted HBO to stop scrambling signals and movie studios to provide first-run movies to pay-per-view operators. They claimed that HBO and other scramblers were trying to kill the TVRO industry.

The HBO reaction to all this came in September 1986 from Michael Fuchs when he said he thought a lot of TVRO dish dealers have been and are still lying to the consumer. He said that the dealers kept telling people that the federal government would soon ban scrambling. In addition, people who have been getting something free for a long time resented having to pay for it now. Smart dish manufacturers were building descramblers into their new dishes. This allowed TVRO owners to call HBO or other pay services on the phone, sign up, and be pulling in a descrambled picture in less than one minute.

Sensing a business opportunity, some cable system operators are getting into the dish-descrambler business themselves. It is not a giant market yet, but it does build up subscriber numbers.

Mek Pandzik, the director of the National Cable TV Cooperative in Overland Park, Kansas, thinks cable operators would have been crazy to get in the TVRO business two years ago and crazy not to get into it today.

Don Shields, Special Projects President at United Cable TV, has put together a $29.95-a-month package of a VideoCipher II descrambler, ten basic, two premium channels, and a guide magazine. Subscribers must take a three-year contract at first, after which they can renew monthly. Extra premium channels are available for $9.95 a month each.

TCI has a similiar package for $29.50 a month for fifteen to twenty basic channels, a leased descrambler, and a choice of two pay services, plus additional pay services for $6.50 a month each. They require a three-year commitment and a one-time $99 charge.

Cable system operators are sometimes surprised to find that TVRO dish owners are emotionally charged up on the scrambling issue. Alan McDonald, Marketing Vice President of Daniels & Associates, warned cable operators that in selling to the TVRO market they were facing a warlike hostility. Dish owners thought of themselves as guerrilla fighters against the entrenched establishment. To overcome such hostility, Daniels & Associates set up its dish-descrambling service as a separate operation from its cable operation. Priscilla Walker is the Director of Consumer Satellite Systems set up by Time Inc.'s ATC to help TVRO dish owners make a smooth transition into the cable system. The basic idea is to bury the hatchet with TVRO dish owners quickly so that everybody can benefit.

Deregulation of the cable industry is another major issue that is already the subject of debate and will continue to be in the years ahead. The cable industry's rates used to be set by local governments, but the Congress abolished that. It deregulated the industry so that back on January 1, 1987, it was freed of local rate regulation and could start charging whatever it wanted to charge.

As some cable operators have a reputation for greed, many

people inside and outside the industry worry about what will happen. Will there be a huge jump in cable rates or what? There have already been some significant rate increases in cable rates. From January 1 to May 1, 1987, subscriber rates jumped an average of 23 percent. Remember that, some years ago, Bob Wussler, an executive with Turner Broadcasting, predicted that one day people would be paying more for cable television each month than they did on their car payments. At the time people thought that Bob's elevator didn't go all the way to the top, but now in 1988 some worry that he might have been right.

Nimrod Kovacs, vice president of marketing with United Cable, believes that deregulation is a marketing opportunity to convince consumers that cable is a very good buy. This also means restructuring programming and making cable simpler for the consumer.

Robert E. Sloss, president of Omega Communications, decided to raise the basic rates on his systems to $12 or $12.95 a month, which is an increase of $3, or a 25 percent increase over when rates were regulated by local government. Some Multiple System Operators (MSOs) such as Daniels & Associates feel they should be getting $15 for basic cable service.

The picture is still not sorted out, but cable operators at the Cable TV Administration and Marketing Society (CTAM) meeting in Boston in 1986 were warned by the Chairman of the National Cable TV Association—Time Inc.'s Trygve Myhren—against raising cable rates "mercilessly." Terry Elkes, Viacom president, urged cable operators not to blow an opportunity of a lifetime by jacking up rates for short-term profits. Cable needs more and better programming, he said, and that is going to cost operators money that they should be prepared to spend.

John Sie, senior vice president, Tele-Communications, Inc., said that programming is the most important challenge facing cable. What is needed, he said, is "punch-through programming" that will increase cable penetration in existing and maturing markets.

In the face of deregulation, Showtime overhauled the rates it charged operators to encourage them to cut retail pay cost to subscribers. In other words, it figured correctly that cable operators would be raising the cost of their basic cable service. So to keep the total rate a subscriber would pay for both basic and a premium channel such as Showtime at the same monthly level, Showtime was cutting its rates. Otherwise, subscribers might cancel the entire cable TV service.

In contrast, HBO gave the biggest profit to the operator who charged the highest price to subscribers. Showtime's idea was to compete more aggressively with HBO. In the past, the industry had been reluctant to lower prices because the premium services such as HBO encouraged higher prices. Also, many operators didn't think the demand for premium services was very price-sensitive.

Disney reworked its rate card so that higher prices didn't mean higher profits to the operator. Its executives believed that this was one of the reasons that Disney was the fastest-growing channel in 1985.

In the face of these changes, HBO tried a different approach several months ago. It set an average subscriber level for each MSO as a floor and demanded that each operator pay that much for HBO regardless of whether the operator's actual number of subscribers fell below that guaranteed floor. On the flip side of that flooring policy, HBO then only charged the operator $1 per month for each new subscriber above that floor level. Thus, the operator had incentive to add new HBO subscribers because of the bigger profit in it for that operator.

This arrangement was to last two years. At the end of the two years, the regular and higher rate would apply to the new subscribers gained over the floor level.

Deregulation notwithstanding, many cable executives realize that local politicians will get them some other way if they appear too greedy. One consultant, Paul Bortz of Browne, Bortz and Coddington, warned that rate raises must be handled very carefully, because if the customer thinks he is getting cheated, he will disconnect. And once a subscriber disconnects, it is very tough to get him back.

Chapter 34

Cable TV's Giant as a Grown-Up

At the beginning of 1988, HBO is a sadder but wiser company. It had endured early years of struggle from 1972 to 1978 followed by the euphoric dizziness of incredible success from 1978 to 1983. In 1984–85, the whole cable business flattened out. Yet, the doomsayers were overstating the HBO case.

For example, the cable subscriptions of HBO have never stopped growing. They are growing more slowly than before, yes, but HBO has never had a year with a drop in subscribers over the previous year. For example, 1986 was the slowest year of growth in HBO's history, and it still netted a hundred thousand new subscribers.

So HBO, while still experiencing lower profits and slower growth, is still the dominant factor in the cable and movie business; HBO was still king of the cable and entertainment mountain, mostly because HBO had to keep up with a changing business or die. For example, in 1986 HBO entered into a joint deal with RCA to build a new satellite, Sat-Com K-3, that would be launched late in the 1980s. The deal underscored the expansionistic philosophy of HBO that has kept it on top. Another change contemplated by Time Inc. was to sell its interest in one of the basic cable channels, USA Network. Time Inc. is a partner with MCA and the Paramount division of Gulf & Western in USA Network (mostly sports and information programming for women) and indicates it is willing to sell its share to its other partners for $40 million.

Time and the other two partners have long argued over the

format of USA Network. Time likes the present narrow focus, but the other two wanted a broader, entertainment-type format. MCA and Paramount also wanted it to get into pay-per-view, and Time objected because then it would compete with HBO and Cinemax.

Another example of forward thinking began in April of 1986, when HBO launched a test of its third pay programming channel, Festival, aimed at a group that had been called "the untouchables" in the past. They were regarded as much tougher to convince that they should pay for a premium channel such as HBO. Because this group was a more difficult sell, HBO and other pay channels previously chose to ignore them and skim the cream of eager customers waiting in line. Now the marketplace had changed, and the untouchables were considered a great pool of potential customers waiting to be tapped.

Getting enough good programs has always been a key concern to HBO as the way to get and keep subscribers. Speaking to the Women in Cable in New York City in June 1986, Jim Mooney, NCTA president, summed up the future of the cable industry with one Jewish word, *khamandaving*. Loosely translated, it means "such things going on that you wouldn't believe."

Most of what is going on in cable is in programming. For example, at the June 16 Cable Forum HBO executives spelled out the programming strategy HBO was adopting in 1986–87 and all the way to the end of the 1980s. It is increasing its emphasis on movies created just for cable TV or, as they are called in cable industry jargon, "made-fors." Even though they had made a multimillion movie deal with Warner Communications two weeks before, HBO executives said that HBO would expand its made-fors under the name HBO Pictures. Rick Bieber, senior vice president of HBO Pictures, said that made-fors scored high in viewer satisfaction. In addition, these movies would be released overseas as theatrical films. HBO planned to increase the number of made-fors in 1987 by 50 percent over 1986, and by the middle of 1987 it should be showing one new made-for a month.

Seth Abraham, senior HBO vice president of programming operations, said that the increase of original programming was especially important for prime time and weekends, which HBO calls The Bread Basket of the pay service. He said that HBO was thinking of beefing up its sports programming and perhaps making offers for part of the National Football League schedule of games.

Cable TV's Giant as a Grown-Up 175

Probably the hottest thing to happen in cable TV in 1986 was the Home Shopping Network. HSN runs two twenty-four-hour channels—HSN-1 and HSN-2—and both offer merchandise for sale by telephone order. Often HSN owns the merchandise itself, having gotten it at a discount from manufacturers and distributors who are overstocked. HSN-1 is received by 8 million homes and deals in modest-priced merchandise, while HSN-2 deals in more expensive merchandise and is received by 2 million homes. HSN pays the cable operator 1 percent to 5 percent of the sales made in the cable operator's service area.

The initial offering of Home Shopping Network was 2 million shares by Merrill Lynch Capital Markets on May 13, 1986. It was to be offered initially at $14–16, but demand was so high it went for $18. However, trading opened on the American Exchange at $42 and closed at $42\frac{5}{8}$ on a value of 2.4 million shares. By May 20, 1986, the price had hit $61\frac{1}{2}$, and by August, it topped at $133 a share. After a 3 for 1 split, it went to $37 a share.

Six other cable home shopping networks were immediately introduced into the marketplace, and their shares had a comparable appreciation. C.O.M.B. operating as Cable Value Network (CVN) jumped 205 percent; Financial News Network (FNN), 160 percent; Tempo Enterprises, 118 percent; Horn & Hardart, 106 percent; Consolidated Stores, 97 percent; Biotech Capital, 88 percent; and Comp-U-Card, 34 percent.

As of June 30, 1986, three of these were actually in operation—HSN, CVN, and Tempo.

What kind of money are we talking about here? HSN's experience is that 8 percent of the subscribers will average fifteen purchases a year for an average price of $36.31. That's $544.65 a year. At this point, HSN has 8 million subscribers and expects to grow to 18 million by mid-1987.

However, Paul Kagan and Lauren R. Rublin of *Barron's Weekly* both note that competition will grow and the novelty will wear off. Taking that into consideration, Kagan scales down the estimated sales per subscriber to $350 per year. The cable operator will get 5 percent, or $17.50.

Not that HSN is making everybody happy. As a hedge against losing its cable access when its current one to three year contracts with cable operators runs out, HSN is using its sudden wealth to make an end-run around the cable operator in some places by buying up UHF-TV stations in the name of one of its companies,

Silver King. By doing that and by dumping off the existing programming on those stations, it is hurting a lot of national TV program syndicators who had contracts to rent programs to those stations.

So, does this mean that with HSN Time Inc. has another competitor for the subscriber's dollar? Yes and no. Yes, it is another competitor for HBO and Cinemax. But Time Inc. is in the home shopping business itself.

HSN's biggest competitor is the Cable Value Network, which is owned by a group of multiple system cable operators, including TCI and Time Inc.'s ATC, the largest and second-largest cable system operators in the country.

As HBO and the cable industry went into 1987, the two biggest financial boosts it had gotten came from the federal government. Washington gave the cable industry a double windfall that took full effect in 1987:

1. **Deregulation.** In the 1984 Cable Communications Policy Act Congress deregulated cable systems' rates completely as of January 1, 1987. That meant that the cable operator could set his own rates.

2. **Killing Must Carry.** In August 1986, the federal courts struck down the FCC's 1962 Must Carry rule, which required cable operators to devote one channel each to every broadcast television station within thirty-five miles. By striking this down, the courts freed up a number of channels for the cable operator, who could then put profit-making programs on them.

In 1988, HBO is still number one in pay cable in America and is continuing to expand. Chastened by two surprising years when the go-go growth stalled, it is rolling once again.

Chapter 35

Eating Broken Glass

My daddy used to say that he who lives by the crystal ball is doomed to eat broken glass. Mindful of that warning, I gingerly approach predictions of where Time Inc., Home Box Office, and the cable industry are headed.

For Time Inc., the glory days are gone. Part of this is due to what happens as corporations get older. In the early phase of an enterprise's life, it is run by the founder, who tends to be charismatic, unpredictable, and passionate and who shares a vision with his employees. That's what Time Inc. was when it was run by its founder, Henry Luce. Driven by a dream of educating the average American about the world, he invented the modern news magazine in 1923 when he began *Time*.

The next phase comes after the passing of the founding father when his corporate heirs try to carry on his passion and dream. That was the administration of Andrew Heiskell, the journalist who headed Time Inc. after Luce left. Then came Dick Munro, the first nonjournalist to take the helm. His arrival marked the third phase of many corporations, namely, the accession of the accounting types, who lack creativity and passion, but who keep track of assets.

Nationally syndicated financial columnist James Flanigan observed in 1986 that the *TV-Cable Week* fiasco marked a watershed event in the history of Time Inc. He said it was proof that Time Inc. had passed from entrepreneurial management to finance-oriented management that avoids risks.

There are three major aspects of what has been going on in Time Inc. during the last decade: corporate confusion, apparent management ineptness, and internal warfare between the two major components of Time Inc.: video and print.

It is the opinion of many Wall Street observers that Time Inc.'s top management clearly has lost its sense of purpose and direction. Not long ago, it hired a management consultant firm to tell it what it was and where it should be going. Nick Nicholas, the new president at Time Inc. as of September 1986, told a conference board meeting of top corporate executives that Time Inc. was trying to figure out what to do next.

The main obsession on the executive floor of the Time-Life building for the last three years seems to have been fending off a hostile corporate takeover and top managers protecting their own jobs. For example, on June 20, 1986, Time Inc. management announced it was going to buy back 16 percent of its own stock to make it tougher for a takeover. Unfortunately, the buy-back plan hasn't worked too well. It was supposed to raise the price of Time Inc. stock to make it more expensive for a corporate raider to buy control, but since the buy-back scheme began, Time Inc. stock dropped in price by over 18 percent, according to a story in *USA Today* on November 4, 1986.

It has not kept Time Inc. from being a potential takeover target. As recently as November of 1986, Coniston Partners, which has been involved in several corporate takeover attempts (Storer Communications and Viacom), reportedly held at least 4 percent and possibly 5 percent of Time Inc.'s stock. News of that immediately sparked Wall Street rumors of another takeover try at Time Inc.

Since 1981, Time Inc. has folded *The Washington Star,* at a loss of $85 million; dumped its subscription TV and teletext experiments, for a loss of $100 million; and, with *TV-Cable Week,* had the biggest magazine flop in history, to the tune of $47 million. A quarter of a billion dollars later, the top brass at Time Inc. responsible for these debacles have been promoted and given dramatic raises while people lower on the totem pole have been put out on the street. On January 30, 1986—labeled Black Thursday—191 editorial and business employees were fired from Time Inc. Demoralized employees complained that the little people were being punished for the mistakes of the top brass. They pointed to the fact that Dick Munro, the number-one executive at Time Inc., was making $583,584 a year when *TV-Cable Week* went under and HBO's profits began to slip. After several failed projects, he is making $712,501.

In fairness, we should note that accountability of top executives doesn't seem fashionable in American corporations. It is as if they are imperial princes who are spectators of what goes on in the companies they are supposedly managing.

Samuel H. Armacost stayed at the helm of the Bank of America as it sank deeper and deeper into the sea of bankruptcy and only resigned when the water began lapping at his shoes. Robert Fomon headed E.F. Hutton during its multimillion-dollar illegal check-overdrafting scheme, but instead of being dumped his board of directors gave him a 25 percent raise.

Charles S. Locke, Mortion Thiokol chairman, never thought about resigning after the Challenger explosion. Warren Anderson, Union Carbide chairman, did offer to resign after Bophal, but his board refused and many executives were surprised at the offer.

In contrast, executives in other countries are expected to be responsible for what their companies do. Jasumoto Takagi, president of Japan Air Lines, personally apologized to the families of the victims of the JAL crash some months ago and Japanese car officials were publicly humiliated when they had to recall one of their models for repairs.

Critics say that Time Inc.'s executive suite is occupied by gray eminences who are obsessed with computer printouts, demographic studies, and all the other paraphernalia of the Harvard MBA. They know little about the vision and passion of the founder, Henry Luce. (Luce never published a magazine that he himself didn't want to read; the present corporate manager never published a magazine he himself *did* want to read.)

Christopher M. Byron, a former editor at Time Inc.'s *Money* magazine and later at the ill-fated *TV-Cable Week,* wrote a book about the latter experience, *The Fanciest Dive.* In the book he describes Time Inc. top management as rife with ambition and power politics and focuses very little on the customer. Uppermost in Time Inc.'s minds is concern about what Wall Street thinks of the last quarter's earnings.

Patrick O'Donnell, a media analyst with Furman, Selz, Meger, Dietz & Burney, has assessed the present corporate stage Time Inc. is in as being just like every other major corporation that is spending a lot of money buying up other companies and trying hard not to be taken over itself.

Financial World's December 11, 1984, story by Robert Sonenclar was entitled, "Despite Its Troubles, Time Still Marches On: But As They Watch the Passing Parade, Some Thoughtful Observers Wonder Whether the Fabled Company Will Ever See the Glory Days Again." Sonenclar said what had happened to Time Inc. in 1983 and 1984 raised some basic questions about how good Time Inc. management was. Certainly, among some Wall Street analysts, Time Inc. has a reputation for bewildering decisions that make about as much sense as if management were relying on reading the entrails of a white angora goat killed during the dark of the moon.

The three men that had formed the key executive team at Time Inc. with Munro were Kelso Sutton, head of the Magazine Group; Nick Nicholas, head of the Video Group; and Jerry Levin, the father of HBO success in pay TV and now the man in charge of long-range planning. In September 1986, Munro moved himself up to chairman of the board and elevated Nick Nicholas to president, thus making Nicholas Munro's heir apparent. In December 1986, Reginald K. Brack, Jr., took over from Sutton at the Magazine (*Money* is a monthly, *Fortune* a biweekly), and a pictorial which is where Brack had been. Both these moves should help. Nicholas is a smart and tough financial taskmaster who will impose, once again, much-needed discipline. Brack has a similar reputation and, as we noted before, helped pull Time-Life Books out of the red. But these men are administrators and not creative types, and the latter may be what is needed at this phase in Time Inc.'s life.

As usual, the top-management inner circle remains the same. Bodies are shifted around from time to time, but they're not being replaced. As one cynic observed, "They ought to be taking to the lifeboats, and instead they are rearranging the deck furniture."

A central internal problem at Time Inc. is the warfare between the video and print divisions. Print had always been the favorite child of Time Inc. and for a long time the only child. In recent years, however, video has taken that role away from print for the simple reason that video—run by young, irreverent Hollywood types—makes the most profit for Time Inc. There is nothing mysterious about it; video makes 46 percent of Time Inc.'s net profit and magazines make 36.5 percent.

In an attempt to reclaim its former status and to give its leader, Kelso Sutton, a shot at the number-one job when Dick Munro retires, the Magazine Group had been going crazy trying to create another smash-hit magazine as it did with *People* magazine. The disastrous *TV-Cable Week* was the result.

TV-Cable Week's failure obviously knocked Sutton out of the running for Munro's job and demoralized the Magazine Group over its ability to successfully create new magazines.

In 1985, Time Inc. bought a regional magazine publisher, Southern Progress Company, for $480 million; *Southern Accents*, which is an interior-decorating publication; and *Asiaweek*. Before it bought these magazines, the print executives spent a year weighing new magazine ideas that they could create in-house. These included a women's magazine (Time had test-marketed and dropped a women's magazine in 1978), a weekly business magazine (*Money* is a monthly, *Fortune* a biweekly), and a pictorial weekly, *Picture Week,* that was ultimately killed after two unsuccessful market tests.

After pondering for about a year, the print executives decided it was safer to buy magazines than create them. So they bought Southern Progress and the others. They later tried to bolster their dropping science magazine, *Discover,* by buying its two main competitors and killing them, but ended up selling *Discover* in 1987 when they couldn't make it successful.

In October 1986, Time Inc. bought another book publisher (it already owns Little, Brown & Company) by the name of Scott, Foresman and Company, the fifth-biggest textbook publisher in the country, for $520 million. This was probably a smart move, since textbooks are a growth business that will be good through the late 1990s and averages better than 15 percent net profit annually. However, the stock market reacted negatively, and Time Inc. stock dropped $3\frac{1}{8}$ points.

At the same time, the Video Group is involved in a $2 billion cable TV system consortium to buy out Group W cable. This move doesn't involve HBO specifically, but it involves the Video Group.

As for new magazines invented in-house, there was *Picture Week,* as we mentioned above, which has been labeled by cynics as the magazine for those who find *People* magazine too intellectual. *Picture Week* fared badly both times it was tried out in thirteen test markets. The diehards in the Magazine Group at Time Inc., desperate to prove they can invent a successful new magazine, urged that *Picture Week* be renamed *Flash* and trotted out for a third market test. Fortunately, wiser heads prevailed and the project was abandoned on November 6, 1986, after $30 million had been spent on it.

In September 1986, Time Inc. also began market-testing another magazine, *Leisure,* which was aimed at the yuppie market. It was also getting involved in a regional, controlled-circulation real-estate magazine and investing $5 million in a joint-venture magazine project called *Parenting,* designed for affluent and educated moms and dads. Beyond that, there was the favorite of Henry Anatole Grunwald, the gentle, obtuse Viennese who has been Time Inc.'s standard showpiece of quality journalism and lifestyle for years. It is an upscale magazine dedicated to the finer things in life and called *Quality.* It came out with one issue in January of 1987, and the new boss at magazines, Reg Brack, decreed it be retired indefinitely. Grunwald followed suit a few months later.

All of what was happening—and not happening—at Time Inc. affected HBO, but HBO also had its own problems. Some of them were of its own making and some were beyond its reach. The problems it made for itself included: The confrontational, combative style of business; blindness to the real world; and young, inexperienced management. The main problem beyond its control was the general market condition of the cable industry.

Some of this I have said before, but in wrapping up our look at HBO and its fortunes, I believe it might be helpful to say it once again for perspective.

HBO is seen in the cable industry and in Hollywood as a heartless and cruel octopus that gives no quarter. I asked *Los Angeles Times* arts critic Charles Champlin how he would characterize HBO. He said, "In just one word: arrogant."

Now HBO wonders why everybody doesn't love it and have a little empathy for its troubles. In truth, the vast majority of people and organizations with whom HBO has done business in the last five years would cheer if it went belly-up. This, of course, it will not do because it is still the giant of pay TV, but that doesn't keep Hollywood people from dreaming.

This mood is reflected in HBO's recent situation. It made a number of long-term movie contracts whose prices were predicated on continued explosive cable subscriber growth. That hasn't happened, and HBO would like to renegotiate its deals with the movie studios.

The movie studios relish the situation and are not inclined to give back any money to HBO on those deals. As one observer said, "The movie studios are eating this up. They still resent the arrogance that HBO displayed in signing these movie deals."

The perceived lessening of HBO's success in 1984–86 was

bad luck, and a result of being unprepared to deal with the changing and maturing pay TV market in spite of plenty of warning from the marketplace.

There were studies, reports, predictions, analyses that all pointed to what was happening, but Home Box Office execs ignored them. As a former Time Inc. Video Group executive described it in the December 11, 1984, *Financial World,* "We never did anything like worst-case planning. We just didn't believe in it. You bet on the high side and believed strongly in your infallibility."

HBO suffered from young, inexperienced management. I believe the average-age employee when I was there in 1983–84 was twenty-seven years old and top management was under forty. No one in top management, incidentally, had ever made a movie, produced a program or TV show, or done anything comparable before coming to HBO.

The young management lacked perspective and maturity. It saw its role as to win, win, win today, and tomorrow be damned. The old corporate phrase, "What goes around comes around," meaning conditions change and he who is on top one day is on the bottom another day, was foreign to top HBO management.

In a way, all of HBO's problems were personified by an event that took place a few days after I joined HBO as chief public relations counsel. HBO executives had planned a reception in Washington, D.C., for members of Congress—to introduce their new comedy series, "Not Necessarily the News," a spoof of politicians.

It was scheduled for a Monday evening in the Corcoran Gallery in downtown Washington, and I was asked what I thought about the plan. I said that most congressmen were not in town on Mondays and Fridays and besides the Corcoran was miles away from Capitol Hill and inconvenient for congressmen. They ignored my objections because they were HBO and believed that everybody would come to see them.

On the morning of the reception, I called the HBO executive offices and said I assumed that the reception had been canceled. When I was told it had not been canceled and asked why it should be, I reminded them that 231 U.S. Marines had been killed by a truck bomb in Beirut the day before. HBO executives said that they knew about that, but what did that have to do with their reception?

I said that no politician was going to show up at a comedy

reception twenty-four hours after 231 U.S. Marines were killed in Beirut. HBO went ahead with it anyhow, and, of course, no congressmen showed up.

By 1985–86, the cable market was running out of new subscribers from which to skim off the cream.

Pay services were being hurt by the impact of the VCR and home video. They fought back by launching advertising campaigns linking cable TV with home video. The theme seen on Showtime, HBO, and other cable channels was that cable was the best friend your VCR has.

A factor helping the VCR was the junk broadcast TV continued to show along with an incredible jumble and clutter of commercials. The number of commercials on broadcast TV in the last two years has jumped from 3,200 to over 5,000. The viewers' disgust is evident in the survey figures.

The August 26, 1986, *Wall Street Journal* reported that advertisers were shifting their dollars away from TV and to direct mail, specialized magazines, and public relations events (such as sponsoring sporting events). The reason was that prime time TV reached 90 percent of American homes in the 1979–80 season, but was only reaching 76 percent in 1985–86.

The corporate chiefs at Time Inc. do seem to be waking up to a new reality. HBO is still the dominant power in movie entertainment, and it will probably continue to be an enormous cash cow for Time Inc. To protect its leading position it has diversified into home video, movie production, theatrical exhibition, and international markets.

Pay cable continues to grow, but 1985 saw its first decline in growth. For example, in 1983 there were 28 million pay cable subscribers, which was 21 percent more than 1982. In 1984, it was up to 31 million, or 10.3 percent higher than 1983. Finally, at the beginning of 1986, it had slipped slightly, to 30.9 million subscribers.

HBO, still the dominant pay service by a wide margin, suffered a 1.3 percent drop in its share of the market from the end of 1984 to the beginning of 1986. At the end of 1984, HBO had 14.4 million, or 46.6 percent of the market. By the start of 1986, it had slipped to 14 million subscribers, or 45.3 percent of the market. By comparison, the next-largest pay service, Showtime, slipped from 17.2 percent of the market in 1984 to 16.6 percent in 1985. By

1987, HBO and Cinemax combined had an estimated 19 million subscribers.

So HBO is almost three times bigger than its nearest competitor. It is the major profit center for Time Inc. and the most dominant player in American entertainment. Yet, one of the most troublesome things for HBO in the months ahead is the uncertainty of the future of its parent, Time Inc.

Time Inc. was expected by financial experts to earn between $3.25 and $3.50 a share in 1986 in operating income, versus $3.15 in 1985. Predictions for 1987 are $3.75 to $4.30 a share. The purchase of Scott, Foresman increases operating income of the Books Group (Time-Life Books, Book-of-the-Month Club, and Little, Brown & Company) to 16 percent of 1985 revenues of $3.4 billion and 18 percent of operating profits of $481 million.

There was a sharp third-quarter rise in net income, to $252 million, or $3.98 a share, up from $44 million, or 70 cents a share, in 1985. But that was mostly not from operating income; operating revenue jumped 7.9 percent to $914 million (up from $847 in 1985).

The main cause of the sharp third-quarter rise was a pretax gain of $352 million from the sale of Time Inc.'s 10 percent holdings in Temple-Inland Company and the public stock sale of 20 percent of ATC. There was also a special one-time charge off of $50 million for moving subscriptions out of Chicago.

Even with all this, profits in three of Time's four operating groups dropped. So, operating income for third quarter 1986 dropped to $98 million from $113 million in 1985.

In the Magazine Group, profits dropped from $24 million last year to $18 million this year because of a drop in advertising and development costs of *Picture Week* and other publications. Books and Information Group profits rose.

Income from HBO and Cinemax dropped from $32 million in the third quarter of 1985 to $24 million in 1986. ATC profits went up slightly, from $27 to $28 million.

David Londoner, media analyst at Time Inc., said earnings were disappointing. "The HBO numbers were worse than anticipated," he said, and added, "The drop in magazine profits was surprising."

There still was some investor displeasure with the purchase of Scott, Foresman. Time stock dropped another $3 to 70\frac{1}{2}$.

Wall Street's view is that Time Inc. is managed by people who are uncertain and confused. This is heard again and again in Wall Street opinions.

While Time Inc. executives routinely dismiss takeover talk, market analysts—such as Gordon Crawford, a senior vice president at Capital Group, Inc., of Los Angeles—see a takeover by Gannett or someone similiar as a distinct possibility. So does John Bauer, an analyst at Oppenheimer & Company who makes the point that the breakup value of Time is $98.28, substantially more than its current stock price of around $71. Paul Kagan expresses a similar view with slightly different figures. Kagan thinks Time Inc. is priced at only about 50 percent of its "inherent value."

ATC has 2.6 million subscribers, which Kagan values at $1,200 each (there have been cable system sales in 1986 and 1987 at an even higher figure), or a total of $3.1 billion. This means, he says, that ATC alone accounts for almost all of Time Inc.'s stock value. Kagan assumes that HBO/Cinemax will have $90 million in cash flow in 1986. With an 8+ multiple (A "multiple" is a number investors give us as a factor by which to multiply the net income of a property in order to get its value. In this case, Kagan is saying to use a factor of something a little over eight. [Multiplying the $90 million cash income of HBO equals Kagan's estimated market value of HBO] of that equals another $750 million in value or $11.77 per share. Add magazines and books with $300 million operating profit (it did $127 million first half of 1985) at multiple of 10 and that equals another $300 million.

Kagan's grand total is $6.75 billion, less $200 million in debt, bringing total Time Inc. value to $6.5 billion. Divide that by 63.7 million shares and it comes to $103 per share. The market price of Time Inc. stock at the close of business on September 18, 1987, was 105.

So, the great media giant, Time Inc., faces uncertain times due in part to its own lack of confidence and painful awareness of management's apparent inability to weave that old magic of years ago when almost everything Time Inc. touched seemed to work. With managers such as Nicholas and Brack in place, Time Inc. will probably heal itself in time. In the past Time Inc. management would stick with a losing magazine such as *Money* or *Sports Illustrated* for years until it finally got the formula for success right. Once it did, the enterprise was golden. However, the economics of the marketplace today are such that the delicious luxury of enough time to do it right is often not there.

What Time Inc. needs now is time. The question is: Has Time run out of time?

The consensus of opinion in the marketplace seems to echo the view of Alan Gottesman, a media analyst for L.F. Rothschild, Unterberg, Towbin, who says, "Time Inc. is a company with a great future behind it."

Epilogue

This book has been several years in the writing, and an enormous amount of research has been done including numerous interviews and analysis of the last twelve years of trade journals, business magazines, business sections of leading newspapers, and a variety of other sources.

The biggest problem of writing about Home Box Office, the Hollywood movie industry, and the world of entertainment is that it is so diversified and so dynamic that something new is breaking every day.

It is not like writing the biography of a dead person, where there is nothing more happening, or the description of an event. This book is not about an event—it is about a process—a process that involves billions of dollars, hundreds of thousands of people in the creative and financial aspects, and hundreds of millions of participants in the audience.

For example, here is a brief rundown of what has happened in the few months since the main part of this book was finished and we were busy editing, typesetting, proofreading and getting ready to print it:

• Frank Biondi and Tony Cox are back together again running HBO's closest rival, Viacom's Showtime.

• Ted Turner announced in January 1987 that he would soon start his own movie channel drawing on the seven hundred or more movies he now owns from the purchase of MGM studios. This movie channel would not be a premium pay channel, but would be advertiser-supported and run as part of the basic cable service.

• The Home Shopping madness continues as the most popular new programming format on cable television. Wall Street cynics first thought it was pure hype and speculation, but changed their views when Sears, Roebuck announced on November 14, 1986, that it had joined the QVC Network to merchandise its products. Following that the December 15, 1986, *Business Week* reported J.C. Penney, K-Mart, and Spiegel were all contemplating jumping in.

Other players getting involved include Rupert Murdoch, Lorimar-Telepictures, and Horn & Hardart, who have combined in a joint venture called "ValueTelevision" (VTV), an hour-long, syndicated home shopping program hosted by Susan Winston.

Then, there is the Crazy Eddie World of Home Entertainment Shopping Network, wherein the New York discount merchandiser goes national. The Sweepstakes Channel will sell discounted magazine subscriptions, books, and audio tapes.

Another one is the American Catalog Shopper's Network (ACSN) done by Mann Media and selling more expensive merchandise drawn from a consortium of several hundred catalog companies. As with most home shopping services, ACSN has an 800 number for customers to call and a sophisticated answering system that can handle up to fifty-seven thousand calls an hour.

The home shopping mix is so eclectic that it also includes TWA airlines and Playboy. Playboy Video is developing what it calls "The Holiday Shopping Show," and will feature furs, lingerie, stereos, electronics, and "other goods associated with the Playboy image," but no sex-related merchandise.

TWA is jumping (or flying) into the marketplace with its Travel Channel, a twenty-four-hour program venture on which it planned to spend up to $15 million in 1987. It will show travel news features and game shows and promote vacation packages, hotels, and resorts. Peter T. McHugh, the president of the Travel Channel, says they don't want to take business away from travel agents, but it seems likely the Channel will, and that might have TWA rethinking the project.

All of this is just the beginning in the home shopping craze on cable. The next step is the shakeout—the consolidation of many home shopping services into a few. The first significant step in that direction came at the end of January 1987 when the pioneer of the business, Home Shopping Network (HSN), announced it

190 *Inside HBO*

was buying out one of its biggest competitors, C.O.M.B., for $646 million. The combined sales of the two for 1987 is projected to be between $1.3 and $1.6 billion.

Cable at the Crossroads in 1987: The January 5, 1987, issue of *Broadcasting* magazine saw the central issue for cable in 1987 as this: "Basic rates are going up and system operators need to develop more programming to help boost viewership and check churn."

Here are excerpts of *Broadcasting*'s analysis, with which a number of industry people agree and all of which has an impact on Home Box Office and its future:[1]

> For years, cable operators have been saying they were in the programming business, not the transmission business. In 1986, they apparently started believing it. In 1987, they may start acting like it.
>
> In a speech before the New York Academy of Television Arts and Sciences last February, National Cable Television Association President Jim Mooney said the industry was on the threshold of "the great age of growth in made-for-cable programming." Ever since then, cable operators have been talking publicly about spending more money on programming to boost penetration and viewership. It's now a question of whether they will go out and do it.
>
> While some may question cable operators' willingness, none should doubt their fiscal ability. A provision of the Cable Communications Policy Act of 1984 that went in effect last week [January 1, 1987] takes state and local governments out of the rate regulation business, freeing operators to raise basic cable rates to whatever level they see fit. And it looks as if most cable operators will be taking advantage of the freedom to raise fees significantly this year—anywhere from 5% to 30%.
>
> Partly to mitigate pay cable churn and partly to soften the impact of the basic rate hikes, many cable operators also plan to lower the price of pay cable this year. But not enough to offset completely the basic rate increases.
>
> Even without any rate increases, cash flow will improve for most MSOs [multiple-system operators] as construction winds down and as systems built in the later 1970s and early 1980s become more efficient and profitable. . . .
>
> So far, the operator's interest in programming has taken the form of buying into cable programming companies. . . . But the promise in 1987 is that the large MSOs will help improve the pro-

[1] Quoted with permission of *Broadcasting* magazine.

gramming schedules of the major basic cable services either by underwriting specific programs or series or by increasing the monthly subscriber fees they pay the services.

Early last summer, TCI [the leading MSO] tried to rally cable operators around the idea of creating a consortium to fund the acquisition or development of "punch-through" programming—drama and sitcoms as good as those on the broadcast networks. Although there was some enthusiasm for the idea early last summer, it seems to be waning. While movie studios and broadcasters may be used to risking millions of dollars for unproved programming in the hope of attracting a substantial audience, cable operators are not.

Putting up millions of dollars to bring improved programming to cable is apparently another matter. Spurred by the talk of a programming superfund, a number of major MSOs have agreed to form a consortium to raise enough money to bid on a package of National Football League games during the 1987–88 season. The consortium would select one of the major basic cable programming services to telecast the games on Sunday nights.

It's far too early to say whether cable will get a fair shot at the NFL. The broadcast networks, which have divvied up the games among themselves, still prize NFL football highly, even though they lost, collectively, more than $125 million on the games during the just-completed season.

Scrambling may well have been the biggest cable story of 1986. On January 15, Home Box Office scrambled full time the feeds of HBO and Cinemax. . . . The scrambling of HBO and Cinemax caused the home satellite industry to go into a tailspin from which it is only now beginning to recover.

The home satellite industry, which was faced with rapidly declining sales, teamed with dish owners, who were faced with having to pay for programming they were used to getting for free, to press Congress for legislation regulating satellite broadcasting last year [1986]. They failed, but they caused Congress to put enough pressure on cable operators to insure that satellite broadcasting would not be dominated by cable operators.

The home satellite market has already shown itself to be an important growth market that can contribute greatly to the [pay cable programmers'] revenue. In 1986, HBO's home satellite sales surpassed all of its projections. At year's end, it counted 90,000 subscriptions, which generated revenue of around $900,000 a month.

The importance of the home satellite market to HBO and Showtime/TMC is greater than it would been a few years ago because of the sluggishness of the cable market. The growth of the pay services started to tail off as new construction wound down and the number of new cable homes started to dwindle. Now pressed hard by competition from home video, pay cable services and their affiliates have had to hustle to sign up new subscribers and discourage old ones from discontinuing service. In 1986, HBO reported a net gain of just 100,000 subscribers, while Showtime acknowledged a net loss of 200,000.

And so the process goes on and whatever way it turns, it seems indisputable that Home Box Office will play a dominant role in the entertainment world of America for a long time to come.

As I said near the beginning of this book, the story of Home Box Office is the story of an almost accidental business that went on to become one of the biggest business successes in America in the last twenty years.

INDEX

Aaron, Dan, 95
Abbey, David S., 41
ABC (American Broadcasting Company), 19, 28, 79–80, 87
  Home View Network of, 55–56
  Nielsen survey on HBO share of households vs., 51–52
  Satellite News Service formed by, 34
Abraham, Seth, 151, 155, 174
Academy of Television Arts and Sciences, 119
ACSN (American Catalog Shopper's Network), 189
Albert, Carl, 23, 26
Alexandria Gazette, 67
Ali, Muhammed, 25
Allen, Herbert, Jr., 165
Allentown, Pa., 7, 11, 14, 84
American Broadcasting Company. *See* ABC.
American Catalog Shopper's Network (ACSN), 189
American Film Theatre, 29
Amira, Sid, 148
Anderson, Warren, 179
Anti-Semitism, 128
Armacost, Samuel H., 179
Arthur, George K., 83
As Summers Die, 161
Asiaweek (magazine), 181
Astoria, Ore., 11
ATC (American Television and Communications), 21, 32, 40, 58, 98, 138, 169, 170, 176, 185
  market analyst's valuation of, 186
Atlanta, Ga., 84
  WTBS in, 27–28
Atlantic Releasing, 162

Babson investment group, 61
Baird, John Logie, 10

Baltimore, Md., 41, 108, 123
Bank of America, 179
Barrington, John, 15
Barron's Weekly, 65, 175
Bauer, John, 186
BBC (British Broadcasting Corporation), 29, 33
Bedell, Sally, 59
Bedell, Thomas, 39
Begelman, David, 44
Belzer, Richard, 160
Benton, Robert, 101
Betamax VCRs, 91, 92
Bethlehem, Pa., 11, 14
Bieber, Rick, 174
Billey, Thomas J., Jr., 169
Billock, John, 57, 125
Biondi, Frank, Jr., 73, 131–36, 163, 188
  as chairman of HBO, 119–21, 127, 135–37
  at Coca-Cola, 136, 148
  as president of HBO, 85, 134
Biotech Capital, 175
Black, Norman, 123
Black Entertainment Network, 156
Blank, Matt, 157
Blay, Andre, 93
Blind bidding, 102
Block booking, 102
Book-of-the-Month Club, 185
Borger, Tex., 88–89
Bortz, Paul, 172
Bouras, James, 153
Boxing. *See under* Sports.
Brack, Reginald K., Jr., 180, 186
Brewin, Bob, 68
Broadcasting (magazine), 97, 112, 138, 156, 190
Bronfman, Edgar, 44
Bronx, The, 94
Brown, Merrill, 119

195

196 Index

Burnett, Carol, 130
Business Week (magazine), 120, 135, 140, 189
Byron, Christopher M., 66, 69, 179

Cable News Network. *See* CNN.
Cable television
  broadcasters' ownership of, 41
  broadcasting industry's attacks on, 19
  California law vs., 18
  churning in. *See* Churning.
  cities still not wired for, 123
  deregulation of, 150, 151, 170–72, 176, 190
  early history of, 11
  early uncertainty about, 3–4, 15
  franchising of, 40–41, 86
  future of, 167–72, 190–92
  home video and, 143–44, 147
  international expansion of, 167
  movie-theater opposition to, 19
  1974 statistics on, 22
  1977 success of, 30
  1979–1982 number of subscribers to, 34, 39–40, 56
  1980–1981 shifts in market of, 56–57
  1981 drop in net profits of, 124
  1982 as bad year for, 60–65
  1984–1985 decline in rates of growth of, 109, 184
  1987 gross income of, 143
  1987 number of homes with, 149
  1987 rise in number of subscribers to, 162
  payment systems for, 11–12
  potential customers for, 151
  predicted growth of, 46
  Premiere anti-trust suit and, 47–48
  promotion by, 156–58
  property owners' rights and, 61–62
  selling price per subscriber in, 55
  split between studios and, 101
  sports events on. *See* Sports—on cable television.
  TV-Cable Week disliked by, 68
  videocassettes/videodiscs and, 41, 143
  Wall Street and, 60, 61, 108
  See also HBO.
Cable Television Administration and Marketing Society (CTAM), 125, 171
Cable Value Network (CVN), 175, 176
Cablevision (company), 9, 84, 156
Cablevision (journal), 41, 95, 135
Caesar's World, 148
Caito, Alan, 98
California, pay TV forbidden by law in, 18
Canadian Film Board, 57
Cannon Group, The, 162
Capital Group, Inc., 186

Capote, Truman, *In Cold Blood*, 52
Carlin, George, 160
CBS (Columbia Broadcasting System), 87
  cable system of, 33, 63–64
  Nielsen survey on HBO share of households vs., 51–52
  Tri-Star Productions and, 77, 100, 134
CBS Broadcast Group, 20
Champlin, Charles, 182
Channels (journal), 92–93
Chaplin, Charlie, 161
Chase, Chevy, 160
Chernin, Peter, 144
Children's Television Workshop, 131
Choice Channel, 146, 148
Churning, 11, 12, 33, 52, 56, 57, 109, 124–26, 141, 158
Chuzmir, Stuart, 15
Cieslak, Susan, 61
Cimino, Michael, 73
Cincinnati, Ohio, 84, 97
Cinemax, 34, 52–53, 120, 139, 155, 157, 158
  movies on HBO and, 109
  1980–1981 growth of, 56–57
  1984 changes in, 107
  "spicy" programming of, 109
Clarke, Arthur C., 22
Cleveland, Ohio, 62, 75
CNN (Cable News Network), 28, 58
Coalition for Better Television, The, 86–87
Coca-Cola, 64, 74, 80, 136, 148
  HBO's 1987 deal with, 162
Collins, Joseph L., 98, 138
Columbia Cablevision, 35
Columbia Studios (Columbia Pictures), 24, 44, 45, 47, 49, 74, 77, 93
  HBO's deals with, 73, 79, 100, 112, 134
  Ghostbusters showing, 160
C.O.M.B., 175, 190
Comcast, 95, 157
Comedy programs, 160
Commercials, 184
Committee to Protect the Public from Paying for What It Now Gets Free, The, 19
Commtron, 93
Communications Daily, 69
Comp-U-Card, 175
Conference Board, The, 108, 141
Coniston Partners, 178
Conley, Ray, 98
Consolidated Stores, 175
Cooke, Jack Kent, 34
Cooney, Gerry, 43, 146
Corwin, Sherrill C., 19
Cox, Tony, 29, 32, 135, 188
Cox Cable, 84, 97, 108, 149
Crawford, Gordon, 186

Index 197

Crook, David, 110
Cruz, Ruben, 85
Crystal, Billy, 160
CTAM (Cable Television Administration and Marketing Society), 125, 171
CVN (Cable Value Network), 175, 176

Daily Variety, 19, 97, 114
Dallas, Tex., 62, 75, 109, 146
Daniels, Bill, 96–97, 146
Daniels & Associates, 55, 84, 96, 170, 171
Davis, Bette, 161
DBA Productions, 106
DBS (Direct Broadcast for Satellite), 167
De Laurentiis Entertainment Group, 162
DeNiro, Robert, 78
Deregulation, 151, 170–72, 176, 190
Derek, Bo, 84
Derick, Chris, 47
Direct Broadcast for Satellite (DBS), 167
Director, Roger, 52
Discover (magazine), 70, 181
Disney, Walt, Studios, 16, 58, 162, 163
Disney Channel, 58, 64, 172
Dodgervision, 146
Dolan, Chuck, 14
  HBO founded by, 6–8, 11–13, 20, 42*n*
  later career of, 9, 34, 84
  Sterling Information Services and, 3, 4
Donaldson, Lufkin & Jenrette, 121
Dooley, Ed, 167
Doubleday, Nelson, 146
Dunn, Polly, 33
Duran, Roberto, 51, 97

Eastwood, Clint, 161
Echo Park, 162
Eden, Lee, 147
El Paso, Tex., 157
Electronic Industries Association, 91
Elkes, Terry, 171
Energy crisis, 16
English, Jim, 153
Entertainment Channel, The (TEC), 33, 41, 58–59
Erlick, E., 19
Escapade Cable Network, 34, 57, 88–89
ESPN (Entertainment and Sports Programming Network), 33, 97
Esquire (magazine), 127, 129, 131
E.T., 105, 114
EventTelevision, 148
Everly, Kateryn, 72
Everly Brothers, 105

Fairchild Group, *Home Video Survey* of, 143
Fairfax County, Va., 61
Fawcett, Farrah, 162

Federal Communications Commission, 17
  on cable industry finances, 124
  cable TV regulation by, 19, 29–31
  HBO's satellite and, 24–25
  movie rentals, 15
  Must Carry rule, 176
  size of receiving dishes, 24, 28
  See also Deregulation.
Festival service, 163–64, 174
Fictionalized nonfiction, 52
Filmways, 134
Financial Analysts Federation, 95
Financial News Network (FNN), 175
Financial World (magazine), 179–80, 183
Finch, Bruce, 51
First Amendment, 85
First National Bank of Boston, 62
FirsTicket, 97
Flanigan, James, 177
"Flashback," 52
Flashpoint, 105
Folies Bergere, 29
Fomon, Robert, 179
Forbes (magazine), 45, 150
Ford, Gerald, 29
Fortune (magazine), 181
Fraggles, The, 105
Frank, Rich, 112
Frazier, Joe, 25
Fuchs, Michael, 29, 32, 90–94, 98, 106, 131, 132, 135, 156, 158, 165, 169
  as chairman of HBO, 14, 127–30, 138, 144
  negotiations with studios by, 48–49, 73
  on PPV, 149–50, 154
  as president of HBO, 127
  Universal deal of, 113–15
Furst, Austin, 29, 67, 94, 111, 130, 133–34

Galkin, Dick, 20
Galuski, Diane, 83
Gannon, Don, 58
Garr, Teri, 105
Getty Oil, 33, 47
Ghostbusters, 160
Giaquinto, Gene, 115
Giera, Walter, 156–57
Gill Cable system, 88
Girard, Tom, 104, 114
Glickman, Marty, 5, 7
Godwin, Ron, 86
Goettel, Gerard, 47–48
Goldblum, Jeff, 161
Goldman, Sachs and Company, 46
Goldwyn, Sam, 102
Goodrich, Henry, 126
Gordon, Sol, 72
Gottesman, Alan, 125, 186

Graham, Jefferson, 73, 85, 100, 113
Granath, Herbert, 79–80
Grand Ole Opry, 58
Green Channel, 4–6
Group W cable system, 58, 64, 97, 104, 157, 181
Grunwald, Henry Anatole, 182
Guccione, Bob, 57

Hackett, Buddy, 160
Hagler, Marvin, 51, 52, 97
Hajdu, David, 112
Hall, Jane, 79
Hall, Rich, 110
Harris, Mel, 129
Hartford, Conn., 18
Hassett, Ann, 106
Hauser, Gus, 108
HBO (Home Box Office)
  as bargain for customer, 151
  chairmen of. See Biondi, Frank; Fuchs, Michael; Levin, Jerry; Nicholas, N. J. "Nick".
  Cinemax of. See Cinemax.
  electronic transmission by, 12
  Festival service of, 163–64, 174
  founding of, 6–8, 11–13, 20, 42n
  as giant entertainment conglomerate, 103, 132, 156, 164
  home video and, 93–94, 143–44, 147, 153, 155
  income of
    first profit, 31
    1973–1975, 26
    1984, 120
    1985–1986, 185
  market analyst's valuation of, 186
  movies on. See Movies—on HBO.
  1983 survey on, 72
  1984 massive firing at, 137–38
  PPV and, 97–98, 150, 154
  Premiere anti-trust lawsuit and, 47–48
  presidents of. See Biondi, Frank; Collins, Joseph L.; Dolan, Chuck; Fuchs, Michael; Heyworth, Jim; Levin, Jerry; Nicholas, N. J. "Nick".
  program mix of, 12–13, 29, 109–10
    1986–1988, 159–66, 174
    original programming, 51–54, 105–6
  promotional efforts by, 157
  recent history of, 182–85
  scrambling by, 168–70, 191
  special projects of, 52, 159
  sports events on. See Sports—on HBO.
  staff outings of, 113
  subscribers of
    churning, 11, 12, 33, 57, 126, 141

  fee structure to cable operators, 172
  free trial period, 157
  Nielsen survey, 107
  1973–1975, 26
  1977, 30
  1980, 33
  1980–1981 market shift, 56–57
  1984 decline, 108–9, 120
  1986 slow growth, 173
  1984–1987, 158, 184–85
  1987 number of, 163
  Take Two service of, 34
  Time Inc. management and, 6, 8, 15, 16, 31, 184–85
  satellite approved, 23
  Time's attempt to sell its cable systems, 20–21
  warning signals to, 123–26, 183
HBO Pictures, 174
HBO Premiere Films, 161
"HBO Showcase '86," 159
"HBO Showcase '87," 159
Hefner, Hugh, 58, 84
Heiskell, Andrew, 25, 77, 177
Henson, Jim, 105
Heyworth, Jim, 15, 24
High Resolution Television, 167
Hills, Morton A., 87
Hirschfield, Alan J., 44, 93
Hoffman, Dustin, 131
Hollywood. See Movie studios; Movies.
Hollywood Reporter, The (periodical), 73, 85, 100, 113
Holmes, Larry, 52, 97, 146, 163
Home entertainment, annual amount spent by Americans for, 142–43
Home shopping networks, 175–76, 189
Home Sports Entertainment (HSE), 146
Home Team Sports (HTS), 146
Home Theatre Network, 88
Home video. See Videocassettes and videodiscs.
Home Video and Cable Report, The, 91
Home Video Publisher (magazine), 143
Home Video Survey (by Fairchild Group), 143
Honolulu, Hawaii, 154
Hooks, Bill, 123
Horn & Hardart, 175, 189
Howard, Kevin, 106
Howard the Duck, 73, 4, 151
HSN (Home Shopping Network), 175, 189–90
Huff, Robert, 169
Hunt, John, 148
Hutton, E. F., 73, 134, 179

Indianapolis, Ind., 62
Intermedia (magazine), 76
Isgur, Lee, 119

J. C. Penny, 189
Jackson, Michael, 92, 94
Jaffe, Barbara, 157
Japan Air Lines, 179
Jenkins, Dan, 128
Johnstown, Pa., 19
Jones, Tommy Lee, 159
Justice, Department of, 29, 47, 79, 100–104, 168–69

Kagan, Paul, 26, 53, 56, 94, 96, 98, 100, 124, 139, 143, 149, 152, 175, 186
Kahn, Irving, 18–19
Kane, Dennis B., 57
Kaufman, Victor A., 77, 80
King, Don, 25, 97, 163
Klein, Robert, 129
Klingensmith, Bob, 111–12
Kovacs, Nimrod, 171
Kristofferson, Kris, 101
KRSC-TV (station), 11
Kulis, Rick, 146

Lambert, Michael, 137
Landro, Laura, 75, 123
Larson, Erik G., 70
Las Vegas, Nev., 148
Lear, William, 4
Leisure (planned magazine), 182
Leonard, Sugar Ray, 51, 106, 146
Levin, Jerry, 5–7, 39, 68, 76, 103, 142, 180
  as chairman of HBO, 29, 44
  as head of Time Inc. Video Group, 32, 44
  as president of HBO, 8, 14–17, 22–29
Lewis, Drew, 109
Lindsay, Robert, 101, 103
Lindstrom, Paul, 151
Little, Brown & Company, 181, 185
"Local access channel," 86
Locke, Charles S., 179
Loew's movie theaters, 156
Londoner, David, 164, 185
Loretto case, 61–62
Lorimar-Telepictures, 189
Los Angeles, Calif., 123, 162
Los Angeles Times, 62, 69, 110, 160, 182
Lowell, Mass., 88
Lower Manhattan. *See* Manhattan Cable TV.
LPTV (Low Power Television), 60
Lucas, George, 72, 135
Luce, Henry, 141, 177, 179

"Made-fors," 174
Madison Square Garden, 24, 45
Mahanoy City, Pa., 11
Mahon, Gigi, 65
Mailer, Norman, *The Executioner's Song*, 52
Maksian, George, 63
Mallardi, Michael P., 79
Malone, John, 150, 154
Mancini, Ray "Boom Boom," 97, 105
Mandel, Howie, 160
Manhattan Cable TV (Sterling Manhattan Cable), 20–21, 29, 86
  restricted access to, 59
Mann Media, 189
Marcovsky, Mike, 96
Margulies, Lee, 160
Marketing & Media Decisions (magazine), 146
Martin, Steve, 129, 160
Matheson, Scott M., 86
Max Headroom, 160
Mayer, Louis B., 102
MCA Inc., 35, 101, 146, 173–74
McCauley, Ian T., 71
McCaw, Elroy, 4
McDonald, Alan, 170
McDonald, Dave, 153
McDowell, Edwin, 69
McGraw-Hill Book Company, 67
McHugh, Peter T., 189
McKinsey and Company, 75
MDS (Multiple Distribution Systems), 60
Media General, 61
Meese, Edwin, III, 89
Meigher, Christopher, 69
Meister, David, 137
Memphis, Tenn., 84
Merger mania, 165
Metropolitan Museum of Art, 57
MGM (Metro-Goldwyn-Mayer), 44, 153, 161, 188
Miami, Fla., 85
Midler, Bette, 29, 129
Miesnieks, Mara, 137, 145
Milwaukee, Wisc., 62, 109
Minnelli, Liza, 159
Monasch, Burton, 80
Money (magazine), 179, 181, 186
Montgomery County, Md., 108
Mooney, Jim, 174, 190
Moral Majority, 86
Morality in Media, 83, 87
Morton Thiokol, 179
Motion Picture Association of America, 144, 153
Movie Channel, The (TMC), 43, 47, 49, 56, 62–63, 77, 79
  See also Showtime/The Movie Channel.

Movie studios
  cable-television domination of, 101–2
    effect of VCRs, 143
  change in nature of, 102
  HBO's financing of films for, 133–34, 160–62
  home video as largest single source of money for, 152
  hostility between HBO and, 13, 15–16, 46–50, 76, 112, 113, 128–29, 182
    truce, 119–22
  new technology always fought by, 13, 119
  1983 cable television's payments to, 106–7
  PPV and, 152–53
  television originally opposed by, 165
Movie theaters
  cable television opposed by, 19
  HBO's ownership of, 156
  movie studios and, 101–3
    split of gross profits, 101
  1987 gross income of, 143
Movies
  budgets of, now include sale to cable television, 101
  on HBO, 158
    exclusivity, 43, 63, 100–101, 112, 114–15, 133–34, 162–63
    financing of movies by HBO, 43, 72–73, 78, 94, 102, 105, 106, 130, 133–34, 164, 174
    first miniseries, 106
    flat-fees, 63
    pre-buys, 78, 115, 130, 133
    R-rated, 84, 89, 164
    shortage, 12, 13, 76–77, 108
    sneak previews, 160
    stockpiling, 53
    studios' reluctance to provide, 15–16, 49–50, 63
  number of, produced annually, 13
  on PPV, 96
  on regular TV stations
    with cuts and commercials, 46
    planned not to be put on pay TV, 19
  release schedules of, 91
  on videocassettes and videodiscs, 91–94
  X-rated, 84, 109
"Movies on Demand," 154
MSOs (Multiple System Operators), 171, 190–91
MTV (Music TV Channel), 62, 92
Multichannel News (newspaper), 67, 108, 124
Munro, J. Richard, 8, 29, 44, 74–75, 80, 140, 177, 178, 180
  TV-Cable Week and, 67–71
Murdoch, Rupert, 164, 189
Music TV Channel. See MTV.

Myers, Bill, 15
Myhren, Trygve, 171

NASA (National Aeronautics and Space Administration), communications satellites of, 21, 22
Nashville, Tenn., 148
Nashville Network, 58
National Cable TV Cooperative, 170
National Federation for Decency in Television, 86
National Geographic Society, 57
Natural, The, 101, 164
NBA (National Basketball Association), 7, 8, 16
NBC (National Broadcasting Company), 28, 33, 41, 87
  Nielsen survey on HBO share of households vs., 51–52
NCTA (National Cable Television Association), 23, 28, 34, 39, 60, 64–65, 84, 121, 168, 169, 171
  CBS Cable's party at, 64
  on theft of programs, 98–99
Nelson, Donald, 40
Nelson, Willie, 101
Nevins, Sheila, 52
New York (magazine), 106
New York Academy of Television Arts and Sciences, 190
New York City, 123. See also Manhattan Cable TV.
New York Daily News, 63
New York Times, The, 34, 59, 61, 62, 69, 71, 90, 94, 101–3, 109, 112, 126, 135, 162
New York Times Cable, 153
News on cable television, 28, 34
  fictionalized, 52
Newsweek (magazine), 120, 140
Newton-John, Olivia, 159
NFL (National Football League), 174, 191
NHL (National Hockey League), 8, 16
Nicholas, N. J. "Nick," 108, 121
  as chairman of HBO, 32
  as head of Time Inc. Video Group, 135
  as president of HBO, 29
  as president of Time Inc., 178, 180, 186
  on Time Inc.'s "soul," 141
Nickelodeon (channel), 62
Nielsen, A. C., Company, 151
  surveys of
    on drop in number of cable viewers, 124
    on HBO subscribers, 107
    on HBO's boxing vs. competing network shows, 51–52
    on home taping by VCRs, 92
    on homes with TV and with cable TV, 149

Norton, Ken, 43
Nosenko, 159
"Not Necessarily the News," 183
NPD Electronic Media Tracking Service, 124
Nudity, 83, 85, 86, 88–89

Oglivy & Mather, 148
Omega Communications, 171
On Cable (magazine), 68
"On Location," 51, 129, 160
O'Donnell, Patrick, 179
ON-TV/Oak, 57
O'Toole, Peter, 161
OPEC (Organization of Petroleum Exporting Countries), 16
Oppenheimer & Company, 186
Opryland Productions, 58
Orion Pictures, 73, 94, 134
Oughton, Carol, 131
Owen, Bruce M., 165

Paley, William, 64
Pandzik, Mek, 170
Paramount Pictures, 16, 47, 49, 63, 77, 79, 100, 111–12, 169
  HBO and, 129, 150, 173–74
  1987 deal, 162
Parenting (planned magazine), 182
Parsons, Ed, 11
Patton, George S., 111
Pay-per-view TV. *See* PPV.
Pay television. *See* Cable television.
Pay TV Newsletter, 94
Penthouse Channel, 57
People (magazine), 32, 69, 180
People's Choice, 147
Personal Preference Video, 169
Philadelphia, Pa., 162
Philadelphia Spectrum Arena, 7
Picture Week (planned magazine), 70, 181, 185
Pirates
  from cable TV, 61, 98–99, 167–70
  of movies, 101
  videocassette, 93
Pirates of Penzance, The, 96, 105
Pittsburgh, Pa., 109
Playboy, 29, 34, 57, 83–84, 88–89, 97, 147
  "The Holiday Shopping Show" of, 189
PPV (pay-per-view) TV, 12, 59, 95–98, 143–54
  cable operator's investment for, 148–49, 150
  expected profit from, 145
  VCRs contrasted to, 146–48
Premiere (cable TV service), 47–50
Premiere Films, 161
Price, Frank, 72, 114

Prime Ticket, 157
Prince, 161
Prism Channel, 162
Private Ticket, 147
Promotion by cable television, 156–58
ProToCall, 157

Quality (planned magazine), 182
Quinones, Ernie, 157

Radio City TV (RCTV), 33
Railroads, the, 133
Rainbow Programming, 34
Rand Corporation, 20
Randolph, Frank, 5
"Ray Bradbury Theater," 161
RCA Corporation
  The Entertainment Channel (TEC) of, 33, 41, 58–59
  satellites of, 24, 173
RCTV (Radio City TV), 33
Reagan, Ronald, 87, 165
Receiving dishes, 23–24, 28
  for one private property, 61
  "pirate" (TVRO), 61, 98–99, 168–70, 191
Redford, Robert, 101
Reiss, Jeffrey, 147
Request Television, 147
Ritchie, Dan, 58
Roberts, Steve, 80
Robertson, Cliff, 44
Robin Hood and the Sorcerer, 105
Rogers, Thomas, 169
Rogers Cable TV, 148, 149
Ronstadt, Linda, 96
Rosenberg, Steven, 143
Rosenstiel, Thomas B., 69, 70
Rosenthal, Sharon, 111
Ross, Steven, 63
Rothschild, L. F., Unterberg, Towbin, 125, 187
Rublin, Lauren R., 175

St. Louis, Mo., 40, 123
Sakharov, 110
Salmans, Sandra, 126
San Antonio, Tex., 148
San Diego, Calif., 97, 149
San Diego Cable Sports Network, 146
San Francisco Examiner, The, 62
San Francisco Giants, 97, 146
San Jose, Calif., 88
Santa Barbara, Calif., 97
Satcom I, 24
Sat-Com K-3, 173
Satellite News, 34, 58
Satellite Television. *See* STV.
Schafly, Phyllis, 86
Scheffer, Steve, 113, 163

Schneider, Jack, 20
Schneider, Roy, 162
Science 86 (magazine), 70
Scientific-Atlanta, 24
Scott, Foresman and Company, 181, 185
Scrambling, 99, 149, 167–71, 191
Sears, Roebuck, 189
SEC (Securities and Exchange Commission), 163
Segal, George, 105
Selleck, Tom, 78
Service Electric Cable TV, Inc., 11
Shales, Tom, 106
Sheinberg, Sidney, 113–15, 164, 166
Shelley, Roger, 87
Shepley, Jim, 15, 16
Sherman Anti-Trust Act, 47
Shields, Don, 170
Shopping networks, 175–76, 189–90
Short, Martin, 160
Showtime, 26, 34, 43, 47–49, 56, 58, 79
Showtime/The Movie Channel, 99, 100–101, 103–4, 107, 111–15
  as bargain for customer, 151
  Biondi and Cox as managers of, 188
  cutting rates after deregulation by, 171–72
  exclusivity policy of, 162–63
  1984–1985 decline of market share of, 184
  1987 number of subscribers of, 162
  Paramount and, 100, 111–12, 129
  PPV and, 143–44, 150
  recent deal by, 162
  as VCR-friendly, 144, 153
  Viewer's Choice of, 147
Sie, John, 171
Silver King, 176
Silver Screen Partners, 73, 134, 161, 163
Silverman, Dave, 21
Simon and Garfunkel, 51
Sinatra, Frank, 129
Sloss, Robert E., 171
SMATV (Satellite Master Antenna Television), 60–62
Smith, Edgar, 4
Smith Barney, Harris, Upham & Company, 137, 145
Smithsonian Institution, 57
Sniglets, 110
Snyder, Ed, 7
Sonenclar, Robert, 179–80
Sony
  sued by the studios, 92
  TeleFirst trial by, 55–56
  VCRs of, 91
Sound Video Unlimited, 93
Southern Accents (magazine), 181
Southern Progress Company, 181

SPACE, the Satellite Association, 168
Space Age Video, 169
Space satellites, 22–27, 173
Spiegel (firm), 189
Spielberg, Steven, 161
Spinks, Leon, 52
Spinks, Michael, 163
Sports
  on cable television
    FCC ruling on, 19
    opposition, 19–20
  on HBO, 8, 16–17
    Ali-Frazier fight, 25–26
    Cooney-Norton fight, 43–44
    later boxing events, 51–52, 105, 106, 129, 163
  on PPV, 96–97, 145–46
Sports Illustrated (magazine), 69, 186
Spotlight, 42, 57, 111–13
Stallone, Sylvester, 78, 115
"Standing Room Only," 51, 129
Staten Island (magazine), 39
Staubach, Roger, 88
Sterett, Karen, 89
Sterling Communications, 3–5, 20
Sterling Information Services, 3, 4, 20
Sterling Manhattan Cable. *See* Manhattan Cable TV.
Stewart, Jimmy, 130
Storer Cable, 39
Streep, Meryl, 78
Streisand, Barbra, 159
STV (Satellite Television), 60, 75, 140
Susskind, David, 77
Sutherland, Donald, 105
Sutton, Kelso F., 68, 70, 135, 140, 180–81
Sweeney, Mac, 169
Sweepstakes Channel, 189

Takagi, Jasumoto, 179
Take Two, 34
Talent Associates, 77
Tauzin, W. J. "Billy," 169
Taylor, Arthur R., 33, 58–59
Taylor, Elizabeth, 130
TCI (Tele-Communications, Inc.), 150, 154, 169, 170, 176, 191
TEC. *See* Entertainment Channel, The.
Tele-Communications, Inc. *See* TCI.
TeleFirst, 55–56
Telemarketing, 157
Telephone, ordering PPV by, 149
TelePrompter, 19, 24, 132
  Westinghouse's purchase of, 34
Telescripts Industries, 149
Telestar, 148
Teletext, 34–35, 67, 70, 140, 178
Television
  commercials on, 184

invention of, 10
movie studios' original attempt to ban, 165
1987 number of homes with, 149
poor quality of, 39
prime time, decline of viewers of, 184
sex and violence on, 28
share of households on HBO vs. commercial networks, 51–52
transmission problems of, 10
See also Cable television.
Television Audience Assessment, Inc., 125–26
Temple, Art, 74
Temple-Inland Company, 66, 185
Tempo Enterprises, 175
Tenten, Bob, 8
Thompson, Tony, 5–8
Thorn EMI, 94, 155
Time Inc.
  book publishers owned by, 181
  Columbia Cablevision interest of, 35
  consulting firm hired by, 75, 178
  Delaware reincorporation of, 74
  earnings of
    future profitability doubted, 125
    1984, 121–22
    1986–1987, 185
    TV as major source, 53, 139, 156
  executive goofs of, 66–67, 178, 180
  future of, 186–87
  market analyst's valuation of, 186
  1986 mass firing at, 178
  recent history of, 177–82
  revision of mission of, 141
  Sterling Communications and, 3–4, 20–21
  as takeover target, 45, 70, 74–75, 108, 140, 178, 186
  teletext system of, 34–35, 67, 70, 140, 178
  TV-Cable Week of, 35, 66–71, 125, 135, 140, 178, 179
  TV subsidiaries of. *See* ATC; Cinemax; HBO; Manhattan Cable TV.
Time Inc. Books Group, 185
Time Inc. Magazine Group, 53–54, 68, 70, 135, 180–82, 185
Time Inc. Video Group, 32, 44, 74, 135, 139, 181
  1981 earnings of, 53
Time-Life Books, 75
Time-Life Cable Communications, 21
Time-Life Films, 33, 44, 49, 67
  home video rights to, 94
Times-Mirror Cable Company, 143
Times-Mirror Company, 42, 112–13
TMC. *See* Movie Channel, The.
Torello, Judy, 110

Trachtenberg, Jeffrey A., 150
Transponders, 22–23
Travel Channel, 189
Tri-Star, 73, 79–80, 100, 101, 104, 134, 161
Tumpowsky, Ira, 148
Turner, Ted, 26–28, 34, 58, 164, 165, 188
Tustin, Calif., 157
TV-Cable Week (magazine), 35, 66–71, 125, 135, 140, 178, 179
TV Guide (magazine), 52, 67, 84, 111
TVRO (Television Receive Only) dishes, 61, 98–99, 168–70
TWA airlines, 189
20th Century–Fox, 44, 47, 49, 79, 164
  HBO's 1986 deal with, 160–61
  videocassette rights sold by, 93

UA (United Artists), 24, 45, 73, 161, 169
UCLA Film School, 57
Ugly George, 86
Union Carbide, 179
Uniontown, Pa., 156
United Artists. *See* UA.
United Cable TV, 170, 171
United Satellite Association, 169
United-Tribune cable company, 108
Universal Studios, 7, 47, 63, 72–73, 77, 79, 100, 105, 146
  HBO deals with, 113–15, 150
  PPV release of, 96
U.S. Supreme Court
  on home VCR recording, 92, 144
  Loretto decision of, 61–62
USA Network, 35, 45, 49, 150, 156, 173–74
USA Today, 98, 178
Utah Association of Women, 85–86

Valenti, Jack, 144
Value Television (VTV), 189
Variety, 96, 104
VCR Theater, 153
VCRs. *See* Videocassette recorders.
Vestron Video, 94
Viacom, 58, 104, 148, 169, 171, 178. *See also* Showtime.
Videocassette recorders (VCRs)
  development of, 91–92
  market for, 151–52
  number of homes with, 143
Videocassettes and videodiscs, 90–94
  ABC program for, 55–56
  competition between pay TV and, 41, 114, 120–21
  copying of copyrighted TV programs by, 35
  growth of use of, 55
  HBO hold on Hollywood broken by, 143

home taping on, 92–93
  from pay channels, 144
  industry profits from, 92
  by movie studios, 152
  1985–1986 gross profits on, 143
  1987 gross income of, 143
  as number-one source of income for studios, 143
  pirate, 93
  PPR and, 146–48
  studios' early release of movies on, 49–50
  of *Wrestlemania I*, 145
View (publication), 79
Viewer's Choice, 147
Village Voice, The, 68
Vogel, Harold, 90
VTV (Value Television), 189

Wagner, Robert, 105
Walker, Priscilla, 170
Wall Street Journal, The, 62, 65, 75, 79, 123, 184, 127
Walson, John, 7–8
Warner, Rod, 39
Warner-Amex Cable, 62–63, 79, 100, 104, 109, 126
Warner Brothers, 63, 77, 79, 100, 134, 102, 104
  HBO's 1986 deal with, 161, 162
Warner Cable, 21
Warner Communications, 62, 97, 104
Warren, Lesley Ann, 161
WASEC (Warner Amex Satellite Entertainment Company), 62
Washington, D.C., 109, 123, 146
  HBO reception for Congressmen in, 183–84
Washington Capitals, 146

Washington Metropolitan Cable Club, 150
Washington Post, The, 90, 91, 106, 119, 135
Washington Star, The, 67, 140, 178
Watson, John, 11
Weaver, Mike, 51
Weaver, Pat, 18
Weinblatt, Mike, 49, 104, 113
Wertheim & Company, 164
Westar I, 23, 24
Western Cable Show, 96, 129
Western Union, 23, 24
Westinghouse Broadcasting, 34, 58, 64
WGN (station), 28
Wildlife Fund, 57
Wildmon, Donald E., 86–87
Wiley, Richard, 25
Wilkes-Barre, Pa., 7–8, 11, 14–15
William Morris theatrical agency, 128, 129
Williams, Robin, 129, 160
Williams, Treat, 162
Winston, Susan, 189
Women in Cable, 174
Woodbury, N.Y., 84
Woodstown, N.J., 29
WOR (station), 28
WPIX (station), 16–17
Wrestlemania I and *II*, 145–46
WTBS (station), 27–28
WTCG (station), 27
Wussler, Bob, 96, 171
Wyman, Thomas, 64, 80

Young and Rubicam, 148

Z Channel, 162
Zahradnik, Rich, 146
Zorthian, Barry, 4, 5
Zucchi, Daniel, 69